SERVING SAFE FOOD

Certification Coursebook

National Restaurant Association

THE EDUCATIONAL FOUNDATION

Disclaimer

The information presented in this reference book has been compiled from sources and documents believed to be reliable and represents the best professional judgment of The Educational Foundation. However, the accuracy of the information presented is not guaranteed, nor is any responsibility assumed or implied by The Educational Foundation of the National Restaurant Association for any damage or loss resulting from inaccuracies or omissions.

Laws may vary greatly by city, county, and state. This book is not intended to provide legal advice or establish standards of reasonable behavior. Operators who develop food safety-related policies and procedures as part of their commitment to employee and customer safety are urged to use the advice and guidance of legal counsel.

Marianne Gajewski, Director, Product Development

Susan M. Myers, Manager, SERVSAFE® Product Development

Virginia A. Christopher, Manager, Production

Beverly E. Sorkin, Associate Editor

Catherine E. Wajer, Assistant Production Editor

Ellen M. Ross, Editorial Assistant

Laura Stone, Cover Art Director

Claudia Parish, Cover Designer

Thomas Armstrong, PhD, writer, Candace Frawley & Associates

SERVSAFE® Serving Safe Food Certification Coursebook

INDUSTRY COUNCIL ON FOOD SAFETY SPONSORS

These sponsors support food safety education and training. Some may provide products, services or tools for food safety training in your operation like posters, newsletters, samples or brochures related to their product offering.

ALASKA SEAFOOD MARKETING INSTITUTE
1200 112th Avenue, N.E.,
Suite C-226
Bellevue, WA 98004
1-800-806-2497

AMERICAN EGG BOARD
1460 Renaissance Drive
Park Ridge, IL 60068
1-847-296-7043

ATKINS TECHNICAL INC.
Thermometer Manufacturer Division
3401 S.W. 40th Boulevard
Gainesville, FL 32608-2399
1-800-284-2842

CAMPBELL SOUP COMPANY
Foodservice Division
Campbell Place
Camden, NJ 08103-1799
1-800-879-7687

COOPER INSTRUMENT CORPORATION
33 Reeds Gap Road
Middlefield, CT 06455-0450
1-800-835-5011
e-mail: cooper@nai.net

DAYDOTS LABEL COMPANY, INC.
2501 Ludelle Street
Fort Worth, TX 76105
1-800-321-3687
website: daydots.com
e-mail: marketing@daydots.com

ECOLAB INC.
Institutional Division
ECOLAB Center
St. Paul, MN 55102
1-800-352-5326

FOODHANDLER, A DIVISION OF ISLAND POLY
514 Grand Boulevard
Westbury, NY 11590
1-800-338-4433

HEINZ U.S.A.
P. O. Box 57
Pittsburgh, PA 15230-0057
1-800-547-8924

HOBART CORPORATION
701 Ridge Avenue
Troy, OH 45374-0001
1-800-960-1190

KATCHALL INDUSTRIES INTERNATIONAL, INC.
5800 Creek Road
Cincinnati, OH 45242
1-800-533-6900
website: katchall.com

KIMBERLY - CLARK
Away From Home Sector
1400 Holcomb Bridge Road
Roswell, GA 30076-2199
1-800-472-6881

LIPTON FOODSERVICE
800 Sylvan Avenue
Englewood Cliffs, NJ 07632
1-800-884-4841
website: lipton.com

NABISCO, INC.,
Food Service Company
7 Campus Drive, P. O. Box 311
Parsippany, NJ 07054-0311
1-800-852-9393

NATIONAL CATTLEMEN'S BEEF ASSOCIATION
444 N. Michigan Avenue
Chicago, IL 60611
1-800-922-2373
website: cowtown.com

NATIONAL PORK PRODUCERS COUNCIL
1776 N. W. 114th Street
Clive, IA 50325
1-515-223-2600

PROCTER & GAMBLE DISTRIBUTING CO.
Two Procter & Gamble Plaza
Cincinnati, OH 45202
1-800-332-7787

RECKITT & COLMAN
Commercial Group
1800 Hammons Tower,
901 St. Louis Street
Springfield, MO 65806-2512
1-800-442-4733

SMITHKLINE BEECHAM PHARMACEUTICALS
111 Hunteres Lane
Devon, PA 19333
1-610-964-8929

SYSCO CORPORATION
1390 Enclave Parkway
Houston, TX 77077
1-713-584-1390

TYSON FOODS, INC.
Food Service
2210 Oaklawn
Springdale, Arkansas 72764
1-800-424-4253
website: tyson.com

US FOODSERVICE, INC.
613 Baltimore Drive
Wilkes-Barre, PA 18702
1-717-831-7500

Acknowledgments

The development of the *SERVSAFE® Serving Safe Food Certification Coursebook* would not have been possible without the expertise of our many advisors and manuscript reviewers. The Educational Foundation of the National Restaurant Association is pleased to thank the following people for the time and effort they dedicated to this project:

Shirley Bohm, MPH, R.S.; Illinois Department of Public Health

Paul Claflin; Bureau of Public Health

Donald E. Cox; Pat and Mike's Restaurant and Bar

Dean Daniel; Oklahoma Restaurant Association

Gary Dixon; Flagstar Companies, Inc.

Gary P. DuBois; Taco Bell Corporation

Larry M. Eils, R.S.; National Automatic Merchandisers Association

Patrick Evans; KFC Corporation

Ralph Fernandez; The Von's Companies

Ernest Gibson, Ph.D; College of DuPage

Steve Grover; The National Restaurant Association

Eric Hicks; KFC Corporation

Philip B. Kirkwood, Jr., R.S.; Michigan State Department of Public Health

George Macht, FMP; College of DuPage

John A. Marcello, R.S.; The Educational Foundation of the National Restaurant Association

Kathy Moorhouse; Foodmaker

Terry Murphy; Wendy's International, Inc.

Marsha Robbins, R.S.; Food Safety Consultant

Edward G. Sherwin, CFE, FMP, Maryland Hospitality Education Foundation

Michael S. Smith; Cracker Barrel Old Country Store

Elmo Smyth; Wisconsin Division of Health

Kevin Tews; McDonald's Corporation

Judith Murray Vetovitz; KFC Corporation

Mary Weber, FMP; Louisiana Restaurant Association

Table of Contents

INTRODUCTION

Serving wholesome, tasty, safe food to your customers is one of your main goals. However, the day-to-day work of running a foodservice operation is complex and demanding. Employees, food, and equipment must be managed and coordinated every minute of every working day.

The Educational Foundation of the National Restaurant Association has designed the *SERVSAFE® Serving Safe Food Certification Coursebook* to help leaders in the foodservice industry meet this challenge. This book will introduce you to ways to prevent food-related illnesses and help you develop and implement a food safety system.

What are your main goals as a leader in the foodservice industry?

1) Protecting people. It is the single most important reason for food safety. If your operation does not handle food carefully, the people you serve may get sick—or worse.

2) Keeping your employees and customers. Food safety is good business. Keeping customers and co-workers safe helps make your restaurant a better place to work and a place where customers return.

3) Preventing food safety errors. Almost any food can become dangerous if not handled safely. Even professional operations—like yours—can make foodhandling errors if they are not careful.

SERVING SAFE FOOD

Food safety depends on every area of your operation working properly—from receiving the food at the loading dock to serving it to customers. You must do your part—but you cannot do it alone. You need to involve all of your employees. Working together for food safety—that is what it is all about.

What is SERVSAFE®?

SERVSAFE is The Educational Foundation's food safety program. It centers on the foodservice leader's role in measuring risks, setting policies, and training and supervising employees.

SERVSAFE includes employee study guides, videos, leader's guide, and other teaching aids for foodservice leaders to use. In addition, those who successfully complete the SERVSAFE course and examination receive a certificate. SERVSAFE training is accepted in most jurisdictions that require training for food safety. For more information on SERVSAFE training, call The Educational Foundation at 312/715–1010 or 800/765–2122.

Format

The *Serving Safe Food* course provides the following materials:

◆ *Serving Safe Food Certification Coursebook*

◆ *Serving Safe Food Trainer's Guide*

◆ *Trainer's Tool Kit:*
   Slides
   Food Safety Showdown!
   Participant Handout Masters

◆ Six videos:
   Introduction to Food Safety
   Personal Hygiene
   Receiving and Storage
   Preparation, Cooking, and Service
   Proper Cleaning and Sanitizing
   Managing Food Safety: A Practical Approach to HACCP

This *Certification Coursebook* has been designed for self-study. You will find exercises at the end of each chapter and an answer key at the back of the book. Complete these exercises before going to your seminar; not only will you understand the course better, it will help you do well on your certification exam. Continue to use your coursebook on the job to review food safety information.

Each food service is unique and must follow company policies and local laws. Throughout the text, space has been provided for you to make notes on policies, laws, and other important points that apply to you and your operation.

Contents of the Certification Coursebook

Part I   *The Challenge to Food Safety*—covers the need for food safety, the hazards that threaten food, and guidelines for training employees in personal hygiene.

Part II   *Developing a Food Safety System*—covers the basics of a Hazard Analysis Critical Control Point (HACCP) food safety system and methods for training employees to run the system.

Part III   *The Flow of Food*—covers methods for purchasing, receiving, storing, preparing, cooking, holding, serving, cooling, and reheating food safely.

Part IV   *Maintaining Sanitary Facilities and Equipment*—covers designing facilities and choosing equipment, cleaning and sanitizing, and controlling pests. In addition, ideas for working with regulatory agencies are included.

The Benefits of Serving Safe Food

A well-designed food safety program protects your employees and customers. Your operation's reputation will be secure. Repeat business from customers and increased job satisfaction among employees can lead to higher profits and better service.

You may benefit directly from reduced or minimized insurance costs. You will most likely benefit by reducing health code violations and becoming less open to lawsuits claiming injury and negligence. In general, serving safe food is vital to your success.

Food Quality

Handling food safely helps preserve its appearance, flavor, texture, consistency, nutritional value, and chemical properties. Food that is stored, prepared, and served properly is more likely to keep its fresh quality.

Profitability

National Restaurant Association figures show that an outbreak of foodborne illness can cost your operation more than $75,000. Cases involving death and serious injury can cost much more. By serving safe food you can avoid:

- ◆ Legal fees.
- ◆ Medical claims.
- ◆ Employees' lost wages.
- ◆ Cleaning and sanitizing costs.

- Loss of food supplies that must be discarded.
- Bad publicity and loss of business and income.
- Being shutdown.

Safe foodhandling can also lead to lower food costs through less waste. Employees' productivity may also improve when they do their jobs right the first time.

Liability

Today, customers are very willing to use lawsuits to obtain compensation for any injuries they feel they have suffered as a result of the food they were served. Under the federal Uniform Commercial Code, a plaintiff bringing a lawsuit must prove that: 1) the food was unfit to be served; 2) the food caused the plaintiff harm; and 3) in serving the food the foodservice operator violated the *warranty of sale* (rules stating how the food in a foodservice operation must be handled).

If the plaintiff wins the lawsuit, he or she can be awarded two types of damages:

1) **Compensatory damages:** awarded for lost work, lost wages, and medical bills the plaintiff may have.

2) **Punitive damages:** awarded in addition to normal compensation to punish the defendant for wanton and willful neglect.

If you have a food safety program in place, you can use a reasonable care defense against a food-related lawsuit. *Reasonable care* is based on proving that your operation had done everything that could be reasonably expected to prevent illness by ensuring that safe food was served. Written standards, procedures, and inspection results are the keys to this defense.

Marketing

Make it clear to your employees and customers that your operation takes food safety very seriously.

Show your employees that:

- Top management is involved in and supports food safety policies.
- Food safety training for managers and all employees is a high priority. Training courses are offered, updated, and evaluated regularly.
- Safe foodhandling procedures are documented, used in regular inspections, and updated as necessary.
- Safe foodhandling is appreciated. Consider awarding certificates for training and giving out small rewards for good food safety records.

◆ You and other upper-level management obey all food safety rules. Set a good example.

Show your customers that:

◆ Your employees know and follow food safety rules. Use employee pins and buttons, place mats, and posters to get across your message. Be sure your employees can answer simple food safety questions from customers.

◆ Customers themselves can help keep food safe. Use your carryout packaging to urge customers to safely handle their food. Station employees near self-serve areas to help customers avoid contaminating food.

Regulation

You must comply with city, county, and state sanitation laws to stay in operation. The health department overseeing your facility usually has the power to levy fines or close any operation that serves unsafe food or has numerous, documented code violations. Therefore, it is in everyone's best interest to work with your local sanitarians.

What a Leader in the Foodservice Industry Needs to Know About Food Safety

The Food and Drug Administration (FDA) recommends that local and state health departments hold the person in charge responsible for knowing and applying the following information:

◆ The diseases that are carried or transmitted by food and the symptoms of these diseases.

◆ Points in the flow of food where hazards can be prevented, eliminated, or reduced and how the procedures meet the requirements of the local code.

◆ The relationship between personal hygiene and the spread of disease, especially concerning cross-contamination, hand contact with ready-to-eat foods, and handwashing.

◆ How to keep injured or ill employees from contaminating food or food-contact surfaces.

◆ The need to control the length of time that potentially hazardous foods remain at temperatures where disease-causing micro-organisms can grow.

◆ Safe cooking temperatures and times for potentially hazardous foods, such as meat, poultry, eggs, and fish.

- ◆ Safe temperatures and times for the safe refrigerated storage, hot holding, cooling, and reheating of potentially hazardous foods.
- ◆ Correct procedures for cleaning and sanitizing utensils and food-contact surfaces of equipment.
- ◆ The types of poisonous and toxic materials used and how to safely store, dispense, use, and dispose of them.
- ◆ The need for equipment that is:
  - — Sufficient in number and capacity.
  - — Properly designed, constructed, located, installed, operated, maintained, and cleaned.
- ◆ The source of the operation's water supply and the importance of keeping it clean and safe.
- ◆ How the operation complies with the principles of a HACCP-based food safety system, as some local health departments require a HACCP system.
- ◆ The rights, responsibilities, and authorities the local code assigns to employees, managers, and the local health department.

The list is a long one! However, as you will see throughout this coursebook, each of these requirements is covered by a common-sense, logical, and complete plan for food safety.

Note: Icons will appear throughout the coursebook to emphasize highly important food safety information. The icons are:

Temperature

Time

Proper Hygiene and Handwashing

Time/Temperature Log

Cleaning and Sanitizing

Observing and Noting

CHAPTER 1: PROVIDING SAFE FOOD

Test Your Food Safety IQ

1. **True or False:** Salmonella is one of the best known food-related illnesses. (See *The Challenge to Food Safety*, page 7.)

2. **True or False:** A food-contact surface is one that touches food or touches a surface that touches food. (See *Cross-contamination*, page 11.)

3. **True or False:** "Clean" and "sanitary" mean the same thing. (See *Clean vs. Sanitary*, page 11.)

4. **True or False:** Soda crackers are a potentially hazardous food. (See *The Foods Most Likely to Become Contaminated*, page 8.)

5. **True or False:** Toxic metals that leech through worn cookware are a physical hazard. (See *How Food Becomes Unsafe*, page 8.)

Learning Objectives

After completing this chapter, you should be able to:

◆ Recognize the challenges to food safety in your operation.

◆ Discuss the main types of contamination.

◆ Identify the foods most likely to become contaminated.

◆ Recognize how food becomes contaminated.

The Challenge to Food Safety

In general, today's American foodservice industry does a good job of serving safe food—but there's room for becoming even better. This industry always faces the threat that the food it serves may endanger employees or customers. The greatest danger is from *foodborne illnesses*, diseases that are carried or transmitted to people by food. Salmonella and Staphylococcus are two of the best known foodborne illnesses. (See *Chapter 2* for information on specific illnesses.)

The Centers for Disease Control and Prevention (CDC) defines an *outbreak* of foodborne illness as an incident in which two or more people experience the same illness after eating the same food. Laboratory analysis must then show that the food is the source of the illness.

As a manager, you face challenges for preventing outbreaks because of the:

◆ Number and types of foods at risk.

◆ Multiple chances for food to become contaminated. Food is at risk at every stage in the *flow of food*, the path from receiving through storing, preparing, cooking, holding, serving, cooling, and reheating that foods in your operation follow.

◆ Type of customer you are serving. Children, elderly people, and people with weakened immune systems are at the greatest risk for foodborne illness. These people are less able to fight off disease and, therefore, are more likely to become ill.

◆ Shortage of trained employees.

The bottom line is that foodservice leaders must carry most of the burden for serving safe food. A good food safety system and a strong training program are vital.

The Foods Most Likely to Become Contaminated

Although any food can be contaminated, the moist, high-protein foods on which bacteria can grow most easily are classified as *potentially hazardous foods*. The U.S. Public Health Service identifies potentially hazardous foods as any foods that consist in whole or part of:

'milk or milk products, shell eggs, meats, poultry, fish, shellfish, edible crustacea (such as shrimp, lobster, crab), baked or boiled potatoes, tofu or other soy-protein foods, garlic-and-oil mixtures, plant foods that have been heat-treated (such as beans), raw seeds and sprouts, sliced melons, and synthetic ingredients (such as textured soy protein in hamburger supplement).'

According to the FDA, *ready-to-eat foods* are properly cooked potentially hazardous foods and raw, washed, cut and whole fruits, and vegetables (including those that had rinds, peels, husks, or shells removed).

How Food Becomes Unsafe
Contamination

Contamination is the unintended presence of harmful substances or micro-organisms in food. There are three main types of hazards (covered in more detail in *Chapter 2*).

◆ **Biological hazards:** Bacteria, viruses, parasites, and fungi. Contamination by bacteria is the greatest threat to food safety.

EXHIBIT 1.1 POTENTIALLY HAZARDOUS FOODS

Milk and Milk Products

Poultry

Fish

Soy-Protein Foods

Tofu

Shell Eggs

Pork

Sprouts

Melons (sliced)

Garlic-and-Oil Mixtures

Beef

Lamb

Shellfish

Cooked Rice, Beans, Potatoes, or Other Heat-Treated Plant Foods

EXHIBIT 1.2 BIOLOGICAL HAZARDS

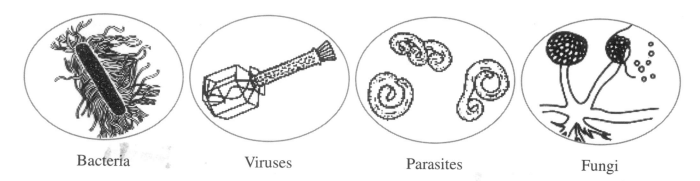

Bacteria　　　　Viruses　　　　Parasites　　　　Fungi

◆ **Chemical hazards:** Pesticides, food additives and preservatives, cleaning supplies, and toxic metals that leach through worn cookware and equipment.

◆ **Physical hazards:** Foreign matter—such as dirt, broken glass and crockery, and other objects—that accidentally get into the food.

EXHIBIT 1.3 CHEMICAL HAZARDS

EXHIBIT 1.4 PHYSICAL HAZARDS

Cross-Contamination

Disease can also be spread by cross-contamination. *Cross-contamination* is the transfer of harmful substances or micro-organisms to food by:

◆ Hands that touch raw foods and then touch cooked or ready-to-eat foods.

◆ Food-contact surfaces that touch raw food, are not cleaned and sanitized, and then touch food that is ready-to-eat.

◆ Cleaning cloths and sponges that touch raw food, equipment, or utensils; are not cleaned and sanitized; and are then used on surfaces, equipment, and utensils for ready-to-eat foods.

◆ Raw or contaminated foods that touch or drip fluids on cooked or ready-to-eat foods.

Food-contact surfaces include any equipment or utensil surface which normally comes in contact with food or which may drain, drip, or splash in food or on surfaces normally in contact with food. Cutting boards, knives, and splash areas are examples of food-contact surfaces.

Clean vs. Sanitary

Clean means free of visible soil. *Sanitary* means free of harmful levels of contamination. Clean food, equipment, and utensils may not be sanitary. For example, a glass may look sparkling clean but may carry harmful bacteria and chemicals. After being washed in boiling water, the same glass may appear cloudy and water-marked, but it is sanitary. For these reasons, the FDA defines sanitization as the use of heat or chemicals to destroy 99.999% of the disease-causing micro-organisms on a food-contact surface.

EXHIBIT 1.5 CLEAN VS. SANITARY

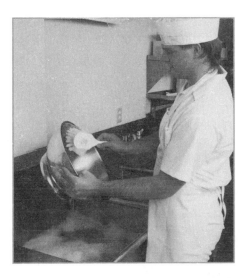

Factors Most Often Named in Foodborne Outbreaks

The following is a list of the most common factors that cause foodborne outbreaks. All these factors need to be controlled when designing your food safety system. Reported cases of foodborne illness usually involve more than one of these factors.

- Failure to properly cool food. (This is the *leading* cause of foodborne outbreaks.)

- Failure to thoroughly heat or cook food.

- Infected employees who practice poor personal hygiene at home and work.

- Preparing food a day or more in advance of being served.

- Adding raw, contaminated ingredients to food that receives no further cooking.

- Allowing foods to stay for too long at temperatures favorable to bacterial growth.

- Failure to reheat cooked foods to temperatures that kill bacteria.

- Cross-contamination of cooked food by raw food, improperly cleaned and sanitized equipment, or employees who mishandle food.

All these factors can be divided into three categories: time and temperature abuse, poor personal hygiene, and cross-contamination.

To help you use the information from this chapter, study *A Case in Point* and see if you can tell what caused the described outbreak of foodborne illness. Then put your knowledge to the test in the *Chapter 1 Exercise*. Answers are found in the *Answer Key* in the back of the coursebook.

A Case in Point

Nearly 200 passengers who had been on an excursion train that traveled through the South experienced foodborne illness on June 14 and 15. At least 55 people went to the hospital. The train had stopped to pick up box lunches prepared by a local restaurant. An analysis of the outbreak implicated the ham in the lunches as the food most likely to have transmitted the micro-organisms that caused the illness.

An investigation showed that on June 11, three days before the lunches were served, 40 hams were delivered to the restaurant and stored in an improperly operating walk-in refrigerator. The next day, June 12, the hams were deboned, cooked, and sliced. Then the hams were cooled but their temperatures were not measured. The slices were refrigerated until the morning of June 14, when the ham portions were boxed with other food for the lunch. The boxes were closed and delivered to the railroad station. The box lunches had been unrefrigerated for three hours before being distributed to the passengers.

All of the ill passengers had eaten lunches containing ham, baked beans, potato salad, rolls, and coffee or tea. A sample from the ham eaten by the passengers was tested, and harmful bacteria in sufficient numbers to cause the illness were identified. A fingernail culture of a foodhandler yielded bacteria identical to that found in the implicated ham.

What do you think happened here?

Answer to a Case in Point

People contributed to this outbreak, and people could have prevented it. In most restaurants, a manager has the greatest responsibility for food safety.

In this instance, one factor was a foodhandler who did not wash his or her hands before slicing the cooked ham. Another factor was inadequate cooling of the ham. The ham also should have been refrigerated because it is a potentially hazardous food. Since the ham slices were not properly cooled, the bacteria kept growing.

Although you might not have been able to guess everything that went wrong in this study, in the following chapters you will be given information to identify individual problems and prevent them from happening.

Chapter 1 Exercise

1. Foodborne illnesses are diseases that are:
   a. carried or transmitted to people by food.
   b. caused by overeating.
   c. cured by proper eating habits.
   d. transmitted to kitchen employees only.

2. Contamination is the:
   a. failure to cook food to the proper temperature to kill bacteria.
   b. mistake of preparing food a day or more in advance of being served.
   c. unintended presence of harmful substances or micro-organisms in food.
   d. accidental mixing of uncomplementary foods.

3. Cross-contamination is the:

   a. main cleaning method for all food-contact surfaces that have been contaminated.

   b. transfer of harmful substances or micro-organisms to food from food or from a nonfood-contact surface, such as equipment, utensils, or hands.

   c. removal of certain bacteria from food by cooking it thoroughly.

   d. prevention of foodborne illnesses.

4. Sanitary means:

   a. free of visible soil.

   b. coated with disinfectant.

   c. washed by a chemical solution.

   d. free of harmful levels of contamination.

5. Moist, high-protein foods on which bacteria can grow most easily are classified as:

   a. potentially hazardous foods.

   b. contaminated.

   c. unfit for children, elderly people, and hospital patients.

   d. requiring pasteurization.

6. Dirt, broken glass, and staples from packaging are classified as:

   a. chemical hazards.

   b. biological hazards.

   c. physical hazards.

   d. bacterial hazards.

7. From receiving through storing, preparing, cooking, holding, serving, cooling, and reheating is called the:

   a. preparation of food.

   b. cycle of food.

   c. activity of food.

   d. flow of food.

8. People who are very young or are already weak or ill are seriously threatened by foodborne illness because they:

   a. cannot take strong medicine.

   b. cannot tell the doctor what is wrong with them.

   c. cannot fight off the disease very well.

   d. are unable to file lawsuits.

CHAPTER 2: FOOD SAFETY HAZARDS

Test Your Food Safety IQ

1. **True or False:** Bacteria can be carried only inside the body of a person or animal. (See *Bacteria*, page 15.)

2. **True or False:** Refrigerating food at 40°F (4.4°C) kills all bacteria. (See *What Bacteria Need to Grow*, page 16.)

3. **True or False:** Bacteria can spread disease after they die. (See *How Bacteria Reproduce*, page 16.)

4. **True or False:** Only allow trained experts to apply pesticides in your restaurant. (See *Pesticides*, page 24.)

5. **True or False:** Store all chemicals in their original containers or transfer them to labeled containers. (See *Foodservice Chemicals*, page 26.)

Learning Objectives

After completing this chapter, you should be able to:

◆ Identify biological, chemical, and physical hazards (dangers).

◆ Describe how bacteria reproduce and grow.

◆ Discuss the factors needed for bacteria to grow.

◆ Know why it is important to keep foods out of the temperature danger zone.

Biological Hazards

Biological hazards are disease-causing micro-organisms, certain plants, and fish that carry *toxins*, which are poisonous. Once in food, some of these hazards may be very hard to kill or control because some are able to survive freezing and high cooking temperatures. Although *micro-organisms* have many good uses, such as ripening cheese and leavening bread, those covered in this chapter cause diseases.

Bacteria

Of all micro-organisms, bacteria are the greatest concern to foodservice managers. *Bacteria* are living single-celled organisms. They can be carried by water, wind, insects, plants, animals, and people. Bacteria survive well on skin and clothes and in human hair, scabs, scars, the mouth, nose, throat, and intestines. Once they contaminate human hands, bacteria may end up in food.

Bacteria may be (see *Foodborne Illnesses Caused by Bacteria* later in this chapter):

◆ *Pathogenic* (infectious, disease-causing). They feed on potentially hazardous foods and can multiply quickly. A disease-causing micro-organism is often referred to as a *pathogen*.

◆ *Toxigenic* (poisonous). These bacteria produce harmful toxins as they multiply, die, and break down; these bacteria are also pathogenic.

Bacteria and the toxins they produce do not have an odor or taste to help you detect them. You cannot tell if they are in food. This is why it is so important to keep them out of food or from growing in the first place.

How Bacteria Reproduce

Bacteria normally exist as *vegetative cells*, which are cells that can grow and reproduce. These cells reproduce by dividing in two (see *Exhibit 2.1*). Each of these cells then divides into two more cells, and so on. As a result, bacteria can multiply to huge numbers very quickly. This rapid rate of reproduction increases the risk of foodborne illness.

Certain bacteria also produce thick-walled protective structures called *spores* inside their cells. Spores may often survive cooking or freezing temperatures and some sanitizing mixtures. Spores do not reproduce, but when conditions around the spores improve, the bacteria become vegetative. This means the bacteria can grow and reproduce.

EXHIBIT 2.1 HOW BACTERIA REPRODUCE

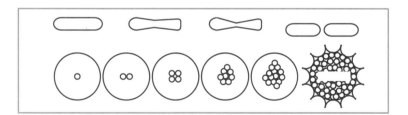

What Bacteria Need to Grow

Bacteria can live anywhere a human can live. In fact, they can often survive extreme temperatures better than people (see *Exhibit 2.2*). Generally, bacteria live well in potentially hazardous foods because these foods are often warm, moist, protein-rich, and neutral or low in acid.

These conditions can be remembered by the acronym FAT-TOM:

F

Food

High-protein foods are likely to be received already contaminated or may be easily contaminated later.

A

Acidity

Acidity is measured on a scale from 0 (very acid) to 14.0 (very alkaline [basic]). A solution with a pH (acid-alkaline measurement) of 7.0 is neutral. Most potentially hazardous foods have a pH level between **4.6 and 7.0** (see *Exhibit 2.3*). However, high acid foods, such as citrus fruit, rarely allow the growth of harmful bacteria. Adding vinegar or lemon juice to food items will help slow bacterial growth—but it does not ensure control and should not be used as the only defense against bacterial growth.

T

Time

Potentially hazardous foods should not remain in the temperature danger zone (see **Temperature**) for more than **four hours** during the foodhandling process.

T

Temperature

The *temperature danger zone** for potentially hazardous foods is **40° to 140°F (4.4° to 60°C)** (see *Exhibit 2.4*). However, since bacteria can survive at (and some bacteria can grow at) lower temperatures, refrigerating food is not total protection against bacterial growth. Discard food if it is past its expiration date.

* The FDA's *1993 Food Code* states that the temperature danger zone is **41° to 140°F (5° to 60°C)**. Some health codes specify **45° to 140°F (7.2° to 60°C)**, while other codes use **40° to 140°F (4.4° to 60°C)** as the temperature danger zone. Check with your local jurisdiction to find out what temperatures are accepted.

O

Oxygen

Some bacteria require oxygen to grow, while others require no oxygen. However, most of the bacteria that cause foodborne illness can grow either with or without oxygen.

M

Moisture

The amount of available water in food is called the *water activity* (a_w). A food with an a_w level of **0.85** or lower is not considered potentially hazardous. Most potentially hazardous foods have water activity values of 0.97–0.99, which is ideal for bacterial growth (see *Exhibit 2.5*). Water activity can be reduced to safer levels by freezing, dehydrating (removing the water), adding sugar or salt, or cooking. Dry foods, such as beans and rice, become potentially hazardous when water is added.

Time

The FDA's *1993 Food Code* states that time only—without temperature—may be used as a bacterial growth control before cooking potentially hazardous foods and for ready-to-eat foods held or displayed for immediate service. To use time as a control, you must:

1. Mark food with the time within which it will be cooked, served, or discarded.

2. Serve or discard food within four hours from the time when it is removed from temperature control.

3. Discard unmarked or time-expired containers or packages of food.

4. Develop written procedures, keep them in your foodservice operation, and make them available to your local sanitarian.

EXHIBIT 2.2 *SALMONELLA* GROWTH CURVE

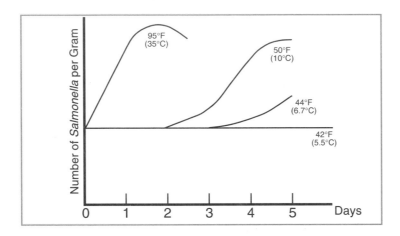

EXHIBIT 2.3 pH OF SOME COMMON FOODS

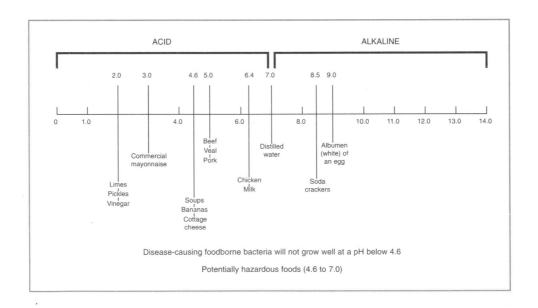

EXHIBIT 2.4 TEMPERATURE DANGER ZONE

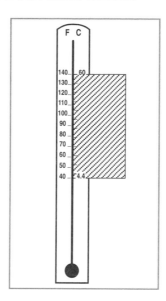

EXHIBIT 2.5 WATER ACTIVITY OF SOME COMMON FOODS

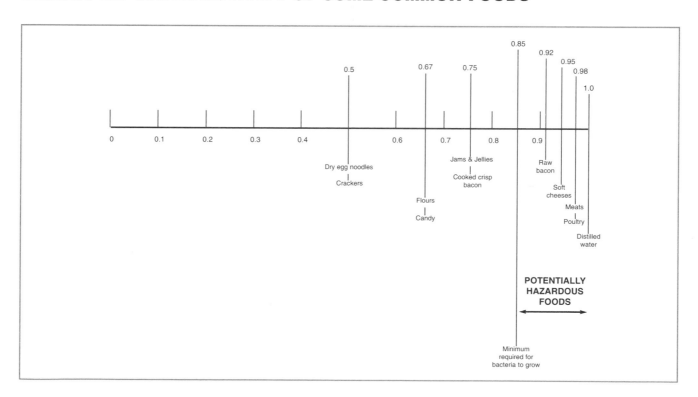

Multiple Barriers To Control Bacterial Growth

As FAT-TOM shows, several common food conditions support bacterial growth. These conditions can occur whenever food is received, stored, thawed, prepared, cooked, held, served, cooled, or reheated.

To control contamination, set up barriers, such as adding lemon juice or prechilling recipe ingredients, that will reduce the conditions of FAT-TOM. The goal is to lower the risk that a single foodhandling error will let bacteria grow enough to cause an illness.

The most important factors to control are temperature and time.

Foodborne Illnesses Caused by Bacteria

Exhibit 2.6 lists the major foodborne illnesses caused by bacteria. To get the most use from this table, review the following terms:

Foodborne infection: A disease that results from eating food containing harmful micro-organisms. Diseases, such as salmonellosis, shigellosis, and listeriosis, are foodborne infections.

Foodborne intoxication: A disease that results from eating food containing toxins from bacteria, molds, or certain plants or animals, such as mushrooms or puffer fish. Staphylococcus and botulism are foodborne intoxications.

Clostridium perfringens, *Bacillus cereus*, and *Escherichia coli 0157:H7* have characteristics of both infections and intoxications.

Incubation period: The time between when an individual consumes contaminated food and when his or her symptoms are present.

Duration of Illness: How long the illness lasts.

Symptoms: The physical signs of the disease.

Source: A host, carrier, or vehicle for disease-causing micro-organisms. A *host* is a person, animal, or plant on which another organism lives and feeds. A *carrier* is a person or animal whose body carries a disease-causing micro-organism. A *vehicle* is an item, such as wind, water, human hands, or dirty utensils, that carries or transports disease-causing micro-organisms.

Foods Involved: Foods that are known to carry or transmit the disease.

Spore Former: States if the bacteria form spores.

Prevention: How the spread of the disease may be stopped.

Viruses

Viruses are protein-wrapped genetic material, the smallest and simplest life-form known. Unlike bacteria, viruses are not complete cells and do not reproduce in food. Like bacteria, however, some viruses may survive cooking or freezing. Also like bacteria, viruses can be transmitted to a human host by food or food-contact surfaces.

EXHIBIT 2.6 MAJOR FOODBORNE ILLNESSES

Disease	Salmonellosis	Shigellosis	Listeriosis	Staphylococcus	Clostridium Perfringens Enteritis
Pathogen	*Salmonella*	*Shigella*	*Listeria monocytogenes*	*Staphylococcus aureus*	*Clostridium perfringens*
Incubation Period	6–48 hours	12–50 hours	few days–3 weeks	rapid	8–22 hours
Duration of Illness	1–2 days (may last longer)	Indefinite, depends on treatment	Indefinite, depends on treatment; high fatality rates in immuno-compromised individuals	2–3 days	24 hours (may last 1–2 weeks)
Symptoms	Abdominal pain, headache, nausea, vomiting, fever, diarrhea	Diarrhea (sometimes bloody), abdominal pain, fever, vomiting, chills, lassitude, dehydration	Nausea, vomiting, diarrhea, headache, persistent fever, chills, backache, meningitis	Nausea, vomiting, abdominal cramps; in more severe cases, headache, muscle cramping, changes in blood pressure and pulse rate	Abdominal pain, diarrhea, dehydration
Source	Domestic and wild animals, humans (intestinal tract)—especially as carriers	Humans (intestinal tract), flies	Soil, water, mud, humans, domestic and wild animals, fowl, damp environments	Humans (skin, hair, nose, throat, infected sores), animals	Humans (intestinal tract), animals, soil
Foods Involved	Poultry and poultry salads, meat and meat products, fish, shrimp, sliced melons, sliced tomatoes, milk, shell eggs, egg custards and sauces, and other protein foods	Salads (potato, tuna, shrimp, chicken, and macaroni), lettuce, raw vegetables, milk and dairy products, poultry, moist and mixed foods	Unpasteurized milk and cheese, ice cream, raw vegetables, poultry and meats, seafood, and prepared, chilled, ready-to-eat foods	Ham and other meats, poultry, warmed-over foods, egg products, milk and dairy products, custards, potato salads, cream-filled pastries, and other protein foods	Cooked meat, meat products, poultry, gravy, beans that have been cooled slowly
Spore Former	No	No	No	No	Yes
Prevention	Avoid cross-contamination, refrigerate food, thoroughly cook poultry to at least 165°F (73.9°C) for 15 seconds (for other foods, see **Exhibit 2.8** for minimum safe internal cooking temperatures), rapidly cool cooked meats and meat products, avoid contamination from foodservice employees by practicing good personal hygiene	Avoid cross-contamination, avoid fecal contamination from foodservice employees by practicing good personal hygiene, use sanitary food and water sources, control flies, rapidly cool foods	Use only pasteurized milk and dairy products, cook foods to proper internal temperatures (see **Exhibit 2.8**), avoid cross-contamination, clean and sanitize surfaces, avoid pooling water	Avoid contamination from bare hands, practice good personal hygiene, exclude foodservice employees with skin infections from food preparation, properly refrigerate food, rapidly cool prepared foods	Use careful time and temperature control in cooling and reheating cooked meat, poultry, and bean dishes and products to 165°F (73.9°C) for at least 15 seconds within two hours
Type of Illness	Infection	Infection	Infection	Intoxication	Toxin-Mediated Infection

EXHIBIT 2.6—continued

Disease	Bacillus Cereus Gastroenteritis	Botulism	Campylobacteriosis	E. coli O157:H7 Enteritis	Norwalk Virus Gastroenteritis
Pathogen	Bacillus cereus	Clostridium botulinum	Campylobacter jejuni	Escherichia coli	Norwalk and Norwalk-like viral agent
Incubation Period	½–6 hours (emetic type); 6–15 hours (diarrheal type)	18–36 hours (may vary from 4 hours–8 days)	2–5 days	2–9 days	24–48 hours
Duration of Illness	Less than 24 hours (emetic); 24 hours (diarrheal)	Several days–a year	7–10 days (relapses common)	8 days	24–60 hours
Symptoms	Nausea and vomiting, occasional abdominal cramps and/or diarrhea (emetic); watery diarrhea, abdominal cramps, pain, nausea (diarrheal)	Lassitude, weakness, vertigo, double vision, difficulty speaking and swallowing, constipation	Diarrhea (watery or bloody), fever, nausea, abdominal pain, nausea, headache, muscle pain	Diarrhea* (watery, could become bloody), severe abdominal cramps and pain, vomiting, occasional low-grade fever	Nausea, vomiting, diarrhea, abdominal pain, headache, low-grade fever
Source	Soil, dust	Soil, water	Domestic and wild animals (intestinal tract)	Animals, particularly cattle, humans (intestinal tract)	Humans (intestinal tract)
Foods Involved	Rice products, starchy foods (potato, pasta, and cheese products); sauces, puddings, soups, casseroles, pastries, salads (emetic); meats, milk, vegetables, fish (diarrheal)	Improperly processing canned low acid foods, garlic-in-oil products, grilled sautéed onions in butter sauce, leftover baked potatoes, stews, meat/poultry loaves	Unpasteurized milk and dairy products, poultry, pork, beef, lamb, non-chlorinated water	Raw and undercooked ground beef, imported cheeses, unpasteurized milk, roast beef, dry salami, apple cider, commercial mayonnaise	Raw shellfish, raw vegetables, salads, prepared salads, water contaminated from human feces
Spore Former	Yes	Yes	No	No	No
Prevention	Use careful time and temperature control and quick-chilling methods to cool hot foods at 140°F (60°C) or higher, reheat leftovers to 165°F (73.9°C) for at least 15 seconds within 2 hours	Do not use home-canned products, use careful time and temperature control for sous vide items and all large, bulky foods, purchase garlic and oil mixtures in small quantities for immediate use and keep refrigerated, cook sautéed onions on request, rapidly cool leftovers	Thoroughly cook food to minimum safe internal temperatures (see **Exhibit 2.8**), avoid cross-contamination	Thoroughly cook ground beef to at least 155°F (68.3°C) for 15 seconds, avoid cross-contamination, avoid fecal contamination from foodservice employees by practicing good personal hygiene	Obtain shellfish from approved, certified sources, avoid fecal contamination from foodservice employees by practicing good personal hygiene, thoroughly cook foods to minimum safe internal temperatures (see **Exhibit 2.8**), use chlorinated water
Type of Illness	Toxin-Mediated Infection	Intoxication	Infection	Toxin-Mediated Infection	Infection

*Other serotypes of E. coli can cause diarrheal illnesses.

Viruses can cause several serious illnesses, such as hepatitis A, which causes inflammation of the liver. These micro-organisms contaminate food through poor hygiene by foodhandlers, contaminated water supplies, or shellfish harvested from sewage-contaminated waters. **The best defense against foodborne viruses is to use good personal hygiene.**

Parasites

Parasites are micro-organisms that need a host to survive. Perhaps the best known parasite is *Trichinella spiralis*, a roundworm found in pigs and certain other game animals. If not killed by thorough cooking or freezing for specified time periods, its larvae can cause *trichinosis*, a disease that causes painful abdominal and muscular cramps. Another parasite is the *Anisakis* roundworm that lives in fish. People who eat raw, marinated, or partly cooked fish may be at risk.

Fungi

Fungi are micro-organisms that range from single-celled plants to mushrooms. Fungi are found in air, soil, and water.

Molds

Single *mold* cells are usually microscopic, but mold colonies may be seen as fuzzy growths on food. The main damage caused by molds is food spoilage, but some molds also produce toxins that can cause illness, infections, and allergic reactions.

Molds can grow on almost any food, at any storage temperature, and under any condition: moist or dry, high or low pH, salty or sweet. Freezing prevents the growth of molds but will not kill those already present in food. The toxins of some molds can withstand cooking. A key food safety control is to throw out foods with molds that are not a natural part of the food. For example, bleu cheese has a natural mold, but many cheeses do not.

Yeasts

Yeasts require sugar and moisture to survive, which they often find in foods such as jellies and honey. Yeasts spoil such products by slowly eating the food. Contamination appears as bubbles, an alcoholic smell or taste, pink discoloration, or slime.

Fish Toxins

The best food safety controls for fish are to buy them only from a reputable and certified supplier and to carefully select the kinds of fish you will serve, noting the following:

◆ Puffer fish, moray eels, and freshwater minnows contain natural toxins.

◆ Certain species of amberjacks, barracuda, and snapper may eat smaller fish that have eaten algae carrying *ciguatoxin*, a naturally occurring toxin. The disease ciguatera is caused by eating fish which through their diet have accumulated elevated levels of ciguatoxin. Symptoms include vomiting, itching, nausea, dizziness, hot and cold flashes, temporary blindness, hot and cold sensory reversal, and sometimes hallucinations. Ciguatoxin is not destroyed by cooking.

◆ Tuna, bluefish, or mackerel that have been kept too long in the temperature danger zone may cause *scombroid intoxication* (from *histamine*, a chemical produced in these temperature-abused fish). Symptoms include flushing and sweating, a burning or peppery taste, nausea, and headache. Other symptoms may include facial rash, hives, edema, diarrhea, and abdominal cramps. Histamine is odorless, tasteless, and not destroyed by cooking. (See *Chapter 6* for information about buying safe fish.)

Plant Toxins

The following plants and foods made from plants have been involved in outbreaks of foodborne illness:

◆ Fava beans, rhubarb leaves, jimson weed, and water hemlock.

◆ Honey from bees that have gathered nectar from mountain laurel, milk from cows that have eaten snakeroot, and jelly made from apricot kernels.

◆ Some varieties of mushrooms. Since poisonous and nonpoisonous mushrooms often look alike, use only mushrooms that are bought from a reliable and approved source. Cooking and freezing do not destroy all plant toxins.

Chemical Hazards

Chemical contamination can be caused by hazards, such as pesticides, food additives and preservatives, cleaning and sanitizing supplies, and toxic metals that leach through worn cookware and equipment. Lubricants used on equipment, personal care products, such as hair sprays, and paints or petroleum products can also contaminate food.

Pesticides

Food safety controls include:

◆ Keep food covered.

◆ Wash all fruits and vegetables before preparation.

◆ Clean and sanitize all equipment and utensils that may have come in contact with any pesticides.

◆ Only allow trained professionals to apply pesticides on your premises.

◆ If you store pesticides, keep them in their original containers. If you put them in a different container, label the new container with the contents and hazards and store them away from food, food-contact surfaces, and other chemicals.

Additives and Preservatives

Sulfiting agents, nitrites, and the smoking process are common preservatives. Monosodium glutamate (MSG) is a common chemical additive. Use only approved preservatives and additives and follow all manufacturers' instructions. *Never* use additives or preservatives to cover spoilage.

Sulfiting Agents

These chemicals are legally used by food processors to preserve freshness and color in certain vegetables, fruits, frozen potatoes and other processed foods, and certain wines. Overuse of sulfites has been linked to a number of serious allergic reactions among sensitive individuals, especially asthma sufferers. Some states forbid restaurants from adding sulfites to food.

Food safety controls include:

◆ Know which of your processed foods and baking items contain sulfites.

◆ If a customer asks about sulfites, tell him or her which menu items contain them. Do not claim foods are sulfite-free unless you know it to be true.

◆ *Never* add sulfites to food.

Monosodium Glutamate (MSG)

This flavor enhancer is on the Generally Recognized As Safe (GRAS) list published by the federal government. However, in some people it may cause flushing, dizziness, headache, dry and burning throat, and nausea. In recipes that do have MSG, use only the called-for amount.

Toxic Metals

Chemical contamination can occur during cooking or storage when certain metals touch high-acid foods. Potentially toxic metals include lead, copper, brass, zinc coating, antimony, and cadmium. Some foods involved in metal poisoning are sauerkraut, tomatoes, fruit gelatins, lemonade, and fruit punches. The following are food safety controls:

◆ Use only food-grade containers.

◆ Use metal and plastic containers and items only for their intended uses. For example, do not use refrigerator shelves that may contain cadmium as makeshift grills or to store unwrapped meat.

◆ Use only proper foodservice brushes on food, *never* wire brushes or ordinary paintbrushes.

◆ Do not use enamelware, which may chip and expose the underlying metal.

◆ Do not allow carbonated water in soft-drink mix systems to flow back into copper water intake lines. The carbonation may leach the copper into the water used to mix drinks (see *Plumbing* in *Chapter 9* for ways to avoid backflow).

◆ Do not use galvanized (zinc-coated) containers for preparing or storing juices, lemonade, tea, or salad dressing.

◆ Do not use lead or lead-based products, including lead-glazed ceramics, in food preparation areas.

Foodservice Chemicals

Detergents, polishes, caustics, cleaning and drying agents, and other similar products are poisonous to humans. Keep them away from food. (See *Chapter 10* for more information about training employees in chemical safety.) The following are food safety controls:

◆ Follow label directions for storing and using chemicals.

◆ Carefully measure chemicals.

◆ Store chemicals in their original containers. Keep them in dry, locked cabinets or areas away from food, food-contact surfaces, and other chemicals that may react with them.

◆ If chemicals are transferred to different, smaller containers or spray bottles, each new container must be properly stored and labeled with the contents and hazards. According to the Occupational Safety and Health Administration (OSHA), all gloves, funnels, measuring cups, and other supplies used to transfer chemicals must also be labeled and stored properly.

- *Never* use food containers to store chemicals or chemical containers to store food. Empty chemical containers must be disposed of as the manufacturer directs.

- Foodhandlers who use chemicals must wash and dry their hands before returning to food preparation duties.

- If you suspect that any food or supplies have been tampered with, label the item and safely set it away from other products, then investigate the situation. Alert your supplier and, if necessary, the health department and the police.

Physical Hazards

Physical hazards include dirt, hair, broken glass and crockery, nails, staples, metal fragments, and other objects that accidentally enter food. The following are food safety controls:

- Do not use glasses to scoop ice. Use only commercial food-grade plastic or metal scoops with handles.

- Do not chill glasses or any food items in ice that will be used for drinks.

- Do not store toothpicks or non-edible garnishes on shelves above food storage or preparation areas.

- Place and maintain protective shields on lights over food storage and preparation areas.

- Clean can openers before and after each use and replace or rotate blade as often as necessary.

- Remove staples, nails, and similar objects from boxes and crates when food is received so these materials do not later fall into the food.

A Case in Point

The Department of Health began an investigation when it received reports that three unrelated patients in the hospital with abdominal cramps and diarrhea that tested positive for *Salmonella*. The investigation showed that one of these patients worked as a cook at a suburban restaurant and that the other two had eaten at that restaurant four days apart. After three days, seven more cases of the illness were reported. The management voluntarily closed the restaurant.

Forty-seven restaurant patrons were interviewed. Twenty-five of them were found to be ill. When questioned, the employees revealed incomplete cooking and poor cooling practices. *Salmonella* was isolated from the cooked prime rib, the cooked roast beef, the cooked ham, the lettuce and the coleslaw, as well as from the surface of a wooden cutting board.

Why do you think the meat was implicated as the cause of this outbreak?

Answer to a Case in Point

Meats are potentially hazardous because they can support the growth of bacteria. Meats are rich in protein and have a pH of about 6.4, the perfect level for bacteria to grow. They also have a high water activity level. The other requirements for bacteria survival and reproduction are oxygen, time, and temperature. Since the meats were not cooked or cooled properly, the bacteria had a perfect temperature and enough time to grow. Cross-contamination of the other foods (the lettuce and coleslaw) happened because the cutting board had not been cleaned and sanitized after the meat had been sliced on it.

In order to limit bacterial growth in potentially hazardous foods, you need to practice careful time and temperature controls. Try to keep foods out of the temperature danger zone during all steps of preparation, storage, serving, and cooling. Also be sure not to contaminate ready-to-eat foods with knives or cutting boards that have come into contact with potentially hazardous foods.

Chapter 2 Exercise

1. The temperature danger zone for potentially hazardous foods is:

    a. 25° to 75°F (–3.9° to 23.9°C).

    b. 85° to 160°F (29.4° to 71.1°C).

    c. 72° to 110°F (22.2° to 43.3°C).

    d. 40° to 140°F (4.4° to 60°C).

2. Bacteria that cause foodborne intoxications:

    a. begin in toxic wastes.

    b. grow more quickly than other bacteria.

    c. produce illness-causing poisons.

    d. can be killed by cooling.

3. Thick walls formed in bacterial cells that can survive some cooking or freezing temperatures and sanitizing solutions are called:

    a. pathogens.

    b. viruses.

    c. parasites.

    d. spores.

4. The two most important factors to control the growth of bacteria are:

    a. oxygen and acidity.

    b. moisture and oxygen.

    c. acidity and moisture.

    d. temperature and time.

5. Bacteria generally grow well in foods that are:

   a. warm, moist, protein-rich, and low in acid.

   b. cool, dry, low in protein, and high in acid.

   c. very hot, wet, calcium-rich, and neutral.

   d. cool, dry, and metallic.

6. The best defense against foodborne viruses is to:

   a. examine food carefully for spores.

   b. use good personal hygiene.

   c. freeze all meat during storage.

   d. blanch all vegetables before preparation.

7. Molds can grow:

   a. only on high-protein foods.

   b. only on high-sugar foods, such as jellies and honey.

   c. on almost any food.

   d. on any foods that do not have a hard skin, casing, or shell.

8. When using pesticides you should:

   a. keep uncovered food in the kitchen.

   b. throw out equipment the pesticides may have touched.

   c. be the person to apply them.

   d. store them away from other chemicals.

9. In choosing cookware:

   a. use enamelware, which never chips.

   b. use galvanized metal containers for storing juices, lemonade, and tea.

   c. use only approved foodservice brushes for food, never ordinary paintbrushes.

   d. use lead or lead-based pans, which do not rust.

10. The best tool for scooping up ice for beverages is:

    a. a sturdy, wide-mouthed water glass.

    b. your hands.

    c. a clean used can or jar.

    d. a commercial food-grade scoop with a handle.

CHAPTER 3: THE SAFE FOODHANDLER

Test Your Food Safety IQ

1. **True or False:** Regular hand lotions help kill bacteria. (See *Sanitizing Lotions*, page 31.)

2. **True or False:** Aprons may be used for wiping hands instead of single-use hand towels. (See *Single-Use Paper Towels and Dryers*, page 31.)

3. **True or False:** Wearing gloves means you do not have to wash your hands. (See *Gloves,* page 33.)

4. **True or False:** People can carry and spread a disease without showing any symptoms of the disease. (See *When an Employee Is Ill or Injured*, page 34.)

5. **True or False:** Smoking in the kitchen is all right if you wash your hands afterwards. (See *Eating and Smoking Areas*, page 35.)

Learning Objectives

After completing this chapter, you should be able to:

◆ Describe the link between personal hygiene and foodborne illness.

◆ Set up basic standards for personal hygiene.

◆ Show support for good personal hygiene by setting an example.

Proper Handwashing

Train your employees to properly wash their hands and make sure that they have proper handwashing stations and supplies.

Handwashing Stations and Supplies

There should be at least one sink set aside *only* for handwashing—*never* to be used for cleaning or for preparing food. Make sure that handwashing sinks also are conveniently located in preparation and warewashing areas. Stations immediately outside restrooms let you observe employee handwashing habits.

Hot-and-Cold Faucets

Each faucet shall allow employees to mix hot and cold water to a temperature of at least 110°F (43.3°C). This temperature is hot enough for proper cleaning but it will not scald (see *Exhibit 3.1*).

Hand Soap

Install dispensers that allow employees to touch only the soap that is given out, not the enclosed supply. Provide nailbrushes to clean fingernails and a sanitizing solution to soak the brushes between uses.

Sanitizing Lotions

Sanitizing lotions or hand dips may be used after washing, but may never be used in place of washing. All lotions must be stored in sealed dispensers. Train employees not to touch food with bare hands until the sanitizing lotion has dried.

Single-Use Paper Towels and Dryers

Hand-drying equipment must be in food preparation areas so employees are not tempted to use their aprons or wiping cloths to dry their hands. Single-use paper towels or air-blowing hand dryers must be provided at each hand sink.

The waste can for used paper towels must be kept clean and set to the side or at a distance from the wash stand. Restrooms used by female employees must include a covered container for sanitary napkins.

EXHIBIT 3.1 PROPER HANDWASHING STATION

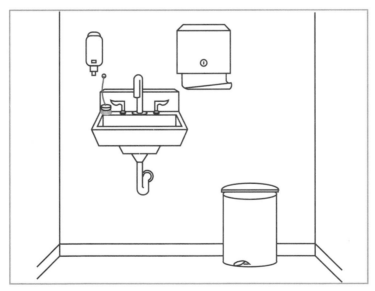

Handwashing

Employees must always thoroughly wash their hands and arms up to the elbow after using the toilet. The FDA's *1993 Food Code* recommends washing hands twice, using a nailbrush as part of the first wash. Employees must also thoroughly wash their hands and arms up to the elbow before starting work and after the following (see *Exhibit 3.2*):

- ◆ Handling raw food.
- ◆ Touching their hair, face, or body.
- ◆ Sneezing or coughing.
- ◆ Smoking and chewing tobacco or gum.
- ◆ Eating or drinking.
- ◆ Cleaning.
- ◆ Taking out the garbage.
- ◆ Touching anything that may contaminate their hands.

After washing their hands, employees should *never*:

- ◆ Use their aprons to dry their hands.
- ◆ Do anything that could recontaminate their hands before returning to work, such as touching their hair.

EXHIBIT 3.2 PROPER HANDWASHING

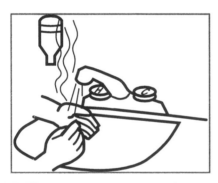

1. Use warm water to moisten hands.

2. Apply soap.

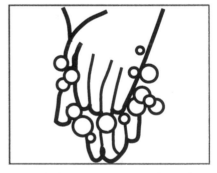

3. Rub hands together for 20 seconds.

4. Rinse thoroughly.

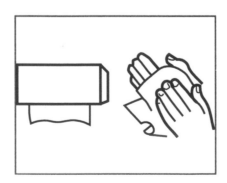

5. Dry.

Hand Care

Basic hand care includes:

◆ Keeping nails short and clean. Not wearing fingernail polish or artificial nails.

◆ Not touching hair, clothes, or skin—especially sores, cuts, or infections.

◆ Covering *all* cuts and sores with bandages and plastic gloves.

Train employees *never* to:

◆ Stack plates to carry several of them at one time—their hands may touch the food.

◆ Handle place settings or food without washing their hands after they have cleared tables or bussed dirty dishes.

◆ Touch the insides of glasses or the eating surfaces of tableware.

Gloves

Employees may use several kinds of gloves—mesh gloves for cutting, rubber gloves for dishwashing, and disposable, plastic gloves for foodhandling. Employees must *always*:

◆ Wash their hands before putting on gloves and when changing into a fresh pair of gloves.

◆ Change gloves as soon as they become soiled or torn and before beginning a different task.

◆ Change gloves at least every four hours during continual use and more frequently when necessary.

Other Rules of Good Employee Personal Hygiene

While personal hygiene may be a sensitive subject, it is vital to food safety. Illness can be spread by almost every part of the human body. Employees should:

◆ Wash their hair and bathe daily.

◆ Wear clean clothing on the job. Work clothes should be worn only on the job, not for personal use. If unable to change clothes at the restaurant, employees should not make any stops on the way to work.

◆ Wear comfortable closed-toed shoes. *Never* wear platform, high-heeled, absorbent-soled, or open-toed shoes.

- Wear hair restraints. These are required by local, state, or federal health codes. Nets, hats, and caps may be used. Employees with beards should also wear beard restraints.

- *Never* wear jewelry. All rings (except for a plain wedding band), bracelets, watches, and necklaces must not be worn while preparing food. They are hard to keep clean and pose a safety hazard if they catch on equipment or accidentally fall into the food.

When an Employee Is Ill or Injured

Employees must report health problems before starting a shift. If they become ill or injured during a shift, they must report it immediately to a manager or supervisor. If an employee's condition may contaminate food or utensils, the employee should stop working for the day and visit a doctor. All medication must be stored with the employee's personal belongings away from food preparation areas.

Bandage cuts, burns, boils, sores, and infections. Bandages should be clean, dry, and prevent leakage from the wound. Waterproof, disposable plastic gloves should be worn over bandages on the hand. Employees wearing bandages may need to be switched to tasks away from food.

Employees with the following conditions should stay at home:

- Fever.

- Diarrhea.

- Upset stomach, nausea, or vomiting.

- Sore throat or sinus infection.

- Coughing or sneezing.

- Dizziness.

Employees need to feel comfortable about talking with a manager or supervisor when they do not feel well. Because:

- People can carry and spread a disease without showing any signs of the disease.

- Even after symptoms disappear, disease-causing micro-organisms can remain in the carrier's body.

Employees may hide an illness to avoid losing pay.

Tasting Food During Preparation

The safest and most sanitary way to taste food is to ladle a small amount of food into a small dish. Taste the food with a clean spoon. Remove the tasting dish and spoon from the area and have them cleaned and sanitized.

Eating and Smoking Areas

Set up eating and smoking areas away from food preparation areas. Many restaurants forbid employees to smoke inside the building.

Storage of Personal Items

As a manager, you should provide a clean, well lighted, secure area where employees can safely store their belongings away from food preparation areas.

Employee Restrooms

These restrooms should be separate from customer restrooms and away from the dining area. Employees should be able to reach them quickly from work areas. If there are not separate restrooms, employees must wash their hands and arms up to the elbow in the restroom and should wash a second time when they return to their work station.

Supporting Good Personal Hygiene

A manager's or supervisor's responsibilities also include reporting certain illnesses to the health department as required by law, scheduling tasks to avoid cross-contamination, and setting a good example with their own personal hygiene practices.

Reporting Illness as Required by Law

Foodservice managers are required by the FDA's *1993 Food Code* to notify health authorities if an employee has or carries the contagious diseases Salmonella typhi, hepatitis A, Shigella, or E. coli O157:H7. Employees may be allowed to continue working under some restrictions, such as wearing gloves or working away from food. On the other hand, employees may be excluded from work for a considerable time or even permanently depending on the disease that they have or carry.

However, a manager or supervisor should not act hastily in excluding employees from work. In some cases, the Americans with Disabilities Act (ADA) and other laws protect employees from being fired or transferred because of illnesses, such as AIDS or testing positive for HIV. The law also protects the confidentiality of employees who report having an illness.

Scheduling Tasks to Avoid Cross-Contamination

Cross-contamination is more likely to happen if the same employee does certain related tasks during the same shift. For example, you should avoid assigning an employee to:

◆ Work with both raw and cooked foods.

◆ Wash dirty dishes and stack clean ones.

◆ Clear dirty dishes and then reset tables with clean dishes.

Setting a Good Example

Follow your operation's procedures for good personal hygiene. Set a good example for employees to follow. Treat all your employees the same. Make sure they all stick to the rules.

A Case in Point

Shigellosis was transmitted to a number of people who went to a Hawaiian Luau Night at a social club. The event had been catered by Chic Catering. A variety of chilled meat salads and fresh fruit and vegetables were served at the party. After the local health department received medical histories from ill party attendees, a sanitarian visited Chic Catering to question the caterer and employees. Keith, the assistant cook and server at the event, reported having had diarrhea on the day of the luau. When asked, the caterer, Jim, said that Keith had made a number of trips to the restroom, but that it had not stopped him from working. Furthermore, Jim said that some of the salads might have been at a room temperature of 70°F (21.1°C) for several hours because they ran out of ice to chill them.

The sanitarian later noticed that the handwashing facilities at the social club were inadequate. Keith confessed that he had not been able to wash his hands thoroughly after each visit to the restroom. Before leaving Chic Catering, the sanitarian also observed that the handwashing station in their kitchen did not have a supply of paper towels and was blocked by several boxes.

What policies could have prevented this outbreak?

What does Jim need to do to correct the sanitation breakdown?

Answer to a Case in Point

Keith did not tell Jim that he was not feeling well. Also, because he did not wash his hands thoroughly after he used the restroom, he contaminated the meat salads and fresh fruit and vegetables with *Shigella*.

Foodservice managers must develop and evenly enforce a policy stating that employees who are ill will not be given foodhandling duties, but will either be reassigned other tasks or sent home until they are well or no longer contagious. Jim should have developed and encouraged open communication with all of his employees so that Keith would have known to stay home when he was sick.

Jim, as the manager, needs to train his employees in the proper personal hygiene habits needed for safe food preparation and handling, as well as telling them the practices that are prohibited. He particularly needs to tell them the importance of handwashing and show them exactly how they need to wash their hands. After training them, Jim needs to monitor and supervise their personal hygiene habits and practices to ensure that they are following employee rules.

Handwashing facilities must be well-equipped and easy for employees to get to as often as necessary. Single-use paper towels or air dryers must *always* be provided. Jim needs to make sure that his employees are able to practice proper handwashing both at job locations and in the kitchen at Chic Catering.

Chapter 3 Exercise

1. Personal hygiene is the:

    a. type of soap and deodorant that a person uses.

    b. number of times a week that a person bathes.

    c. complete medical record of a person.

    d. way a person maintains their health, appearance, and cleanliness.

2. The most important rule of foodservice personal hygiene is that employees must:

    a. wear gloves at all times.

    b. completely give up smoking.

    c. wash their hands often.

    d. see a doctor twice a year.

3. Employees who wear disposable gloves should:

    a. avoid washing their hands before putting on gloves.

    b. wash their hands before putting on gloves.

    c. wash off the gloves if they become soiled with food.

    d. apply hand lotion before putting on gloves.

4. Handwashing stations should allow employees to:

    a. conveniently wash their hands when necessary.

    b. wash and dry their hands in 10 seconds.

    c. stand up straight while washing their hands.

    d. use the facilities without extensive training.

5. When employees wash their hands, they should also wash their:

   a. gloves.

   b. elbows.

   c. face.

   (d.) lower arms up to the elbow.

6. After washing their hands, employees should avoid:

   a. putting on gloves.

   b. talking to other co-workers.

   (c.) touching their hair.

   d. walking through the kitchen.

7. Aprons should be:

   (a.) changed when dirty and never used as towels.

   b. used as towels after handwashing.

   c. washed after no more than two shifts.

   d. always worn home after work.

8. Employees should wash their hands in the sink used for:

   a. thawing foods.

   b. rinsing vegetables.

   (c.) handwashing only.

   d. washing dishes.

9. The safest way to taste food while preparing it is to:

   a. taste directly from the ladle used for stirring.

   b. have another employee hold the spoon while you taste.

   c. use a large spoon, then rinse it so it can be used again.

   (d.) ladle a small amount of food into a dish and taste it with a clean spoon.

10. A manager is required by law to report if an employee has a:

   a. broken leg.

   b. common head cold.

   c. small cut that is properly bandaged.

   (d.) contagious disease that may be spread when working with food.

CHAPTER 4: INTRODUCING THE HACCP SYSTEM

Test Your Food Safety IQ

1. **True or False:** All recipes require flowcharts. (See *Identifying CCPs*, page 42.)

2. **True or False:** Receiving is a critical control point (CCP) for all foods. (See *Identifying CCPs*, page 42.)

3. **True or False:** Only managers should monitor critical control points (CCPs). (See *Monitoring CCPs*, page 46.)

4. **True or False:** Employees need to memorize all HACCP terms. (See *Training Your Employees*, page 48.)

5. **True or False:** Good performance in food safety should be tied to hiring, promotions, and raises. (See *General Food Safety Training*, page 48.)

Learning Objectives

After completing this chapter, you should be able to:

◆ Describe the main principles of a HACCP system.

◆ Assess food safety hazards.

◆ Identify critical control points (CCPs).

◆ Set up procedures and standards for critical control points (CCPs).

◆ Monitor critical control points (CCPs).

◆ Take corrective actions.

◆ Set up a record-keeping system.

◆ Verify that your system is working.

◆ Manage a HACCP system.

◆ Train employees to follow HACCP procedures.

What is HACCP?

A *Hazard Analysis Critical Control Point (HACCP)* food safety system helps you:

◆ Identify the foods and procedures that are most likely to cause foodborne illness.

◆ Build in procedures that reduce the risks of foodborne outbreaks.

◆ Monitor all procedures to ensure food safety.

Common HACCP Terms

Hazards are:

1. Micro-organisms that can grow during preparation, storage, and/or holding.

2. Micro-organisms or toxins that can survive heating.

3. Chemicals that can contaminate food or food-contact surfaces.

4. Physical objects that accidentally enter food.

Risks are the chance that a condition or set of conditions will lead to a hazard.

Critical control point (CCP) is an operation (practice, preparation step, or procedure) where a preventive or control measure can be applied that would:

1. Eliminate (remove) a hazard.

2. Prevent a hazard.

3. Lessen the risk that a hazard will happen.

Steps to Building a HACCP System

The following steps will help you start your HACCP system.

Step 1: Assessing Hazards

1.1 Identifying Potentially Hazardous Foods

Review your menu items and recipes (see *Exhibit 4.1*). Remember that a potentially hazardous food may be served alone or as an ingredient in a recipe. For example, chicken is often served as an ingredient in soup and as a main entree.

1.2 Flow of Food

The *flow of food* is the path food travels in your restaurant.

◆ Receiving.

◆ Storing.

◆ Preparing.

◆ Cooking.

◆ Holding.

◆ Serving.

◆ Cooling.

◆ Reheating.

EXHIBIT 4.1 CHILI RECIPE

Ingredients	Weights and Measures
Ground beef (pre-cooked)	5 lb, 13 oz
Chili base	1 can
Small red beans	1 can
Dark red kidney beans	1 can
Vegetables (frozen)	2 packages
Seasoning	1 packet
Water	1¼ gal

PREPARING

1. Drain and rinse dark red and small red kidney beans. Set aside.
2. Pour chili base into stock pot. Add water and seasoning. Stir with wire whisk until all seasoning is dissolved.

COOKING

3. Preheat stove. Begin heating chili mix.
4. Break up any clumps in the frozen vegetables. Add to the chili mix. Stir with long-handled spoon.
5. Add cooked ground beef and stir. Continue heating chili until 165°F (73.9°C) or higher is reached for at least 15 seconds.

SERVING AND HOLDING

6. Serve immediately.
7. Hold chili at 140°F (60°C) or higher. Do not mix new product with old.

COOLING

8. Cool in shallow pans with a product depth not to exceed 2 inches. Product temperature must reach 40°F (4.4°C) or lower within 4 hours. Stir frequently.
9. Store at a product temperature of 40°F (4.4°C) or lower in a refrigerated unit. Cover.

REHEATING

10. Reheat chili to a product temperature of 165°F (73.9°C) or higher for at least 15 seconds within 2 hours—one time only.

SANITATION INSTRUCTIONS: Measure all temperatures with a cleaned and sanitized thermocouple or thermometer. Wash hands before handling food, after handling raw foods, and after any interruption that may contaminate hands. Wash, rinse, and sanitize all equipment and utensils before and after use. Return all ingredients to refrigerated storage if preparation is interrupted.

1.3 Identifying Hazards

Now that you have set aside the menu items and recipes that use potentially hazardous foods, you need to decide what hazards can occur during the flow of food. Watch your employees in action. Ask them for facts on how temperatures are measured and recorded. Review your current records for any information on how menu items are handled. Make notes about possible hazards for use in your flowchart. Remember your number one worry is bacterial contamination and growth.

1.4 Estimating Risks

Several factors can increase the chance of foodborne illness.

1. **Type of Customers.** Consider who your customers are. Children, elderly people, and customers with weakened immune systems may have a lower resistance to foodborne illness. Plan your menu and recipes to protect all your customers.

2. **Suppliers.** You need reputable (and in some cases, certified) suppliers for all potentially hazardous foods, especially for those that will not be cooked.

3. **Size and Type of Operation.** You need proper equipment and facilities if you plan to serve multi-step recipes. Can your equipment maintain proper temperatures? If you find a recipe is too hard to handle safely, consider replacing potentially hazardous ingredients, making the steps simpler, or buying pre-prepared items from a reputable source.

4. **Employees.** Employees need training to handle food safely. See *Exhibit 4.4* under Standard (Criteria) for examples of hazards.

Step 2: Identifying CCPs

Identify the CCPs needed to keep each recipe safe. Add these to each written recipe (see *Exhibit 4.2*), flowchart (see *Exhibit 4.3*), and written system (see *Exhibit 4.4*). CCPs differ for each food and method of preparation. Although in each recipe CCPs are not needed at every stage in the flow of food, they are needed in one or more stages.

For example, raw chicken may be delivered carrying *Salmonella*—even if it is received at the proper temperature. At receiving, standard procedures are developed to measure the temperature and check for any visual signs of a problem. However, the risk of *Salmonella* can be eliminated during the cooking process by heating the chicken to 165°F (73.9°C) or higher for at least 15 seconds. This makes cooking a CCP.

EXHIBIT 4.2 CHILI RECIPE WITH CCPS

Ingredients	Weights and Measures
Ground beef (pre-cooked)	5 lb, 13 oz
Chili base	1 can
Small red beans	1 can
Dark red kidney beans	1 can
Vegetables (frozen)	2 packages
Seasoning	1 packet
Water	1¼ gal

PREPARING

1. Drain and rinse dark red and small red kidney beans. Set aside.
2. Pour chili base into stock pot. Add water and seasoning. Stir with wire whisk until all seasoning is dissolved.

COOKING

3. Preheat stove. Begin heating chili mix.
4. Break up any clumps in the frozen vegetables. Add to the chili mix. Stir with long-handled spoon.

CCP 5. Add cooked ground beef and stir. **Continue heating chili until 165°F (73.9°C) or higher is reached for at least 15 seconds.**

SERVING AND HOLDING

6. Serve immediately.

CCP 7. **Hold chili at 140°F (60°C) or higher.** Do not mix new product with old.

COOLING

CCP 8. Cool in shallow pans with a product depth not to exceed 2 inches. **Product temperature must reach 40°F (4.4°C) or lower within 4 hours.** Stir frequently.

9. Store at a product temperature of 40°F (4.4°C) or lower in a refrigerated unit. Cover.

REHEATING

CCP 10. **Reheat chili to a product temperature of 165°F (73.9°C) or higher for at least 15 seconds within 2 hours—one time only.**

SANITATION INSTRUCTIONS: Measure all temperatures with a cleaned and sanitized thermocouple or thermometer. Wash hands before handling food, after handling raw foods, and after any interruption that may contaminate hands. Wash, rinse, and sanitize all equipment and utensils before and after use. Return all ingredients to refrigerated storage if preparation is interrupted.

CCPs are highlighted in boldface.

EXHIBIT 4.3 CHILI FLOWCHART

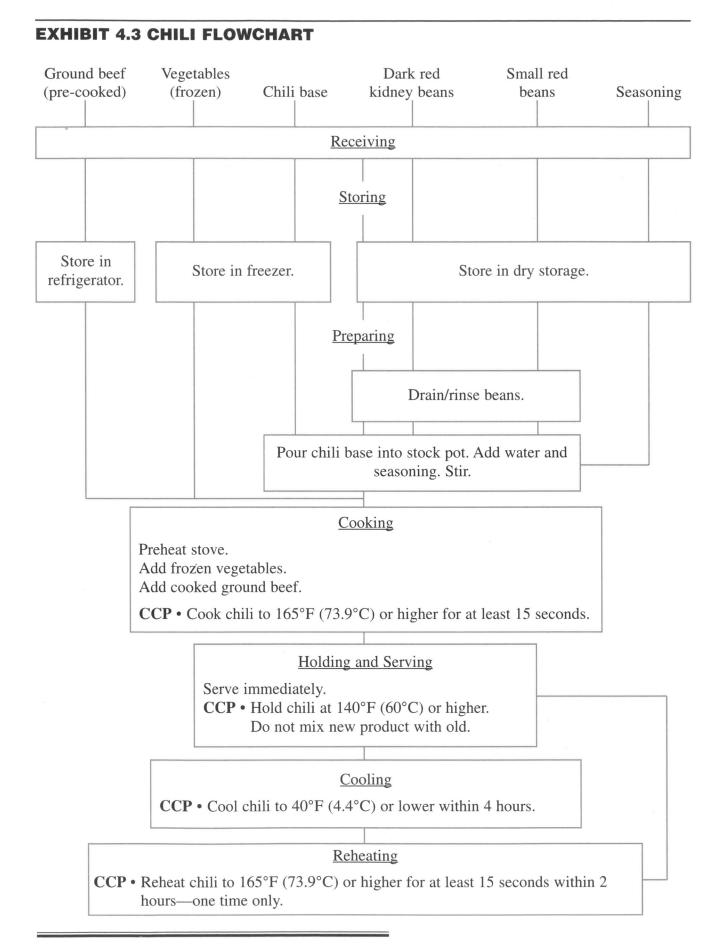

Ground beef (pre-cooked)   Vegetables (frozen)   Chili base   Dark red kidney beans   Small red beans   Seasoning

Receiving

Storing

Store in refrigerator.   Store in freezer.   Store in dry storage.

Preparing

Drain/rinse beans.

Pour chili base into stock pot. Add water and seasoning. Stir.

Cooking

Preheat stove.
Add frozen vegetables.
Add cooked ground beef.

CCP • Cook chili to 165°F (73.9°C) or higher for at least 15 seconds.

Holding and Serving

Serve immediately.
CCP • Hold chili at 140°F (60°C) or higher.
Do not mix new product with old.

Cooling

CCP • Cool chili to 40°F (4.4°C) or lower within 4 hours.

Reheating

CCP • Reheat chili to 165°F (73.9°C) or higher for at least 15 seconds within 2 hours—one time only.

To pinpoint when you need to add a CCP, you might consider designing a flowchart for your recipe.

Designing flowcharts

A *flowchart* is a simple diagram showing the flow of food and all of your CCPs. It is a way to picture what happens to a recipe's ingredients through receiving, storing, preparing, cooking, holding, serving, cooling, and reheating.

Although you will need to create flowcharts for quite a few menu items, this task may not take as much time and effort as you think it will. Similar recipes call for similar flowcharts. For example, the recipes and flowcharts for chicken noodle soup and beef noodle soup will be very similar. Beef roast and baked ham will also be similar.

After you have added the CCPs to your flowchart, add them to your recipe. Your recipe now may be used for guiding the preparation of the menu item.

Step 3: Setting Up Procedures and Standards for CCPs

Set the standards that must be met at each CCP. *Standards* (also termed critical limits by the FDA) are times, temperatures, or other requirements that must be met to keep a food item safe. Add these standards to your written recipe and flowchart.

You may need more than one standard at each CCP. No matter how many you need, each standard should be:

◆ Measurable.

◆ Based on facts from experience, suppliers' advice, research data, or food regulations.

◆ Right for the recipe when prepared in a normal work environment, considering room temperature, number of employees, and number of orders.

◆ A clear direction to take a specific action, such as measuring a temperature or cooking an item for a certain length of time. For example, a standard for reheating chili should be "Heat rapidly on stove to an internal product temperature of 165°F (73.9°C) or higher for at least 15 seconds within 2 hours."

Beside standards for CCPs, write in standards to prevent contamination at other points in the recipe and the flow of food. For example, "Wash, rinse, and sanitize all equipment and utensils before and after use."

Step 4: Monitoring CCPs

Monitoring is checking to see if your standards are being met. Having CCPs without monitoring defeats the purpose of your food safety system. To monitor, you should:

◆ Focus on CCPs throughout the flow of food.

◆ Decide if your standards are being met.

◆ Make sure employees are involved in this process, understand the CCPs, and know your standards.

Step 5: Taking Corrective Actions

When you find a standard for a CCP is not being met, **correct it right away**. Many corrective actions are very simple, such as continuing to heat an item if the end cooking temperature has not been reached. Other corrective actions may not be as simple, such as throwing out a food item. You may want an employee to ask a supervisor before taking action.

To work well, corrective actions must meet the criteria for standards in **Step 3**—they must be based on facts, for normal working conditions, and have measurable goals.

For example, the standard for holding baked chicken may read:

◆ "Hold baked chicken at 140°F (60°C) or higher until served."

The corrective action if the standard is not met may read:

◆ "If held over two hours, discard. If held less than two hours and temperature falls below 140°F (60°C), reheat to 165°F (73.9°C) or higher for at least 15 seconds—one time only."

Step 6: Setting Up a Record-Keeping System

Records should be simple and easy for employees to use. Some ideas are to keep:

◆ Blank forms and a clipboard near work areas to check several items at the same time.

◆ Notebooks to write down what actions have been taken.

◆ All flowcharts and recipes near work areas, so employees can use them quickly.

◆ Blank forms for temperatures hung on equipment for easy use.

If records are easy to use, your employees are less likely to *dry lab*, which is to record data *without* actually measuring the food's temperature.

Step 7: Verifying that the System is Working

Verifying is proving that your system is working. This step usually follows after you have developed your written system. Once you have all of your procedures in place and have decided where your CCPs are, you need to follow the flow of food to make sure what you have decided is correct. Verify you have:

◆ Listed procedures in order.

◆ Identified and assessed all hazards.

◆ Selected CCPs.

◆ Set standards.

◆ Selected monitoring procedures and schedules.

◆ Developed corrective actions.

◆ Decided on procedures and forms for recording data.

◆ Set up procedures to make sure that monitoring is done properly.

◆ Noted any flaws or omissions in procedures.

◆ Calibrated monitoring equipment.

Typical Challenges to a HACCP System Over Time

A HACCP system must be kept up-to-date. For example, you may need to revise the system when:

◆ Changes in customers, suppliers, equipment, or facilities create new hazards or make some of your standards or corrective actions invalid.

◆ Menus and recipes are changed.

When you have completed the written recipe and HACCP system for recipes using potentially hazardous foods, keep them in notebooks so employees can easily reach and use them.

The Written HACCP System

An illustration of what a HACCP system might look like in written form would include operational steps, hazards, CCPs, standards, types of monitoring, corrective actions, and records (see *Exhibit 4.4*). Your local regulatory agency may also require a written HACCP system that includes the food safety principles discussed in this chapter.

The written procedures you have developed in Steps 1–7 provide you with a complete, flexible framework so your system can change as needed.

Your next important step is to train your employees according to the new procedures.

Training Your Employees

Your food safety training program may already cover much of the information employees will need to run a HACCP system. Key goals in adapting your program to support a HACCP system are to:

- ◆ Help your employees understand the basics of a HACCP system. When you first put in the new system, your employees' greatest concern will be how it will affect the work they do. Reassure them that they are already using many of the right procedures. Let them know their role in putting the system to work. Openly talk with employees, so they understand not only what they must do, but feel free to ask questions.

- ◆ Discuss CCP monitoring procedures and recordkeeping.

- ◆ Help employees adjust their current skills to HACCP methods.

- ◆ Identify areas where employees lack knowledge or skills and design training to meet those needs.

Overall, training should be as practical as possible. Employees do not need to memorize the complete HACCP system and its terms. They need to understand the official food safety procedures that directly relate to their jobs.

General Food Safety Training

Before employees can be trained in using HACCP procedures, they need to have a basic understanding of food safety. This information includes:

- ◆ Benefits of practicing food safety.

- ◆ Which foods are potentially hazardous (see *Chapter 1*).

- ◆ How contamination can occur at any point in the flow of food.

- ◆ Time and temperature standards for potentially hazardous foods and the temperature danger zone.

- ◆ Links between good personal hygiene and food safety (see *Chapter 3*).

- ◆ Employees' role in preventing cross-contamination (see *Chapter 1*) and in preventing foodborne illness.

- ◆ Food safety is a requirement in hiring, promotions, raises, and bonuses and should be directly incorporated in performance appraisals.

EXHIBIT 4.4 HACCP SYSTEM FOR CHILI

Step: Purchasing and Receiving

OPERATIONAL STEP	HAZARD	CCP	STANDARD (CRITERIA)	TYPE OF MONITORING	CORRECTIVE ACTION IF STANDARD NOT MET	RECORDS
Purchasing and receiving frozen ground beef	Bacterial growth and survival; chemical or physical contamination		Product obtained from an approved source	Shift manager checks purchase specifications upon receipt of product and invoice	Reject delivery; obtain product from an approved source	Certificate of conformance and invoice
			Accept product at 0°F (−17.8°C) or lower	Shift manager measures temperature with a thermocouple or thermometer	Reject delivery	Receiving log
			Packaging intact	Observation	Reject delivery	
Receiving frozen vegetables	Contamination and spoilage		Accept product at 0°F (−17.8°C) or lower	Shift manager checks for signs of thawing and ice crystals	Reject delivery	
			Packaging intact	Observation	Reject delivery	
Receiving cans of chili base	Contamination		Cans sealed properly, undented, no signs of rust, seams intact and no signs of bulging	Observation and inspection of a random number of cans	Reject delivery	

EXHIBIT 4.4—continued

Step: **Storing**

OPERATIONAL STEP	HAZARD	CCP	STANDARD (CRITERIA)	TYPE OF MONITORING	CORRECTIVE ACTION IF STANDARD NOT MET	RECORDS
Storing frozen ground beef	Bacterial growth and survival		Store frozen patties in freezer at a unit temperature of 0°F (−17.8°C) or lower	Shift manager measures frozen product temperatures with a thermocouple or thermometer	Move to freezer unit capable of maintaining temperature at 0°F (−17.8°C) or lower until ready to thaw or prepare	Daily inspection audit
			Label, date, and use FIFO method of stock rotation	Check that product is covered and packaging intact	Discard if maximum storage time is exceeded	
Storing frozen vegetables	Contamination		Store frozen vegetables in freezer at a unit temperature of 0°F (−17.8°C) or lower	Shift manager checks for signs of thawing and ice crystals	Move to freezer unit capable of maintaining temperature at 0°F (−17.8°C) or lower until ready to thaw or prepare	Daily inspection audit
			Label, date, and use FIFO method of stock rotation	Check that product is covered and packaging intact	Discard if maximum storage time is exceeded	
Storing canned chili base			Store cans in dry storage. Label, date, and use FIFO method of stock rotation	Observation	Discard if maximum storage time is exceeded	Daily inspection audit

EXHIBIT 4.4—continued

Step: Thawing

OPERATIONAL STEP	HAZARD	CCP	STANDARD (CRITERIA)	TYPE OF MONITORING	CORRECTIVE ACTION IF STANDARD NOT MET	RECORDS
Thawing frozen ground beef	Bacterial survival and growth		Keep product at or below 40°F (4.4°C) during thawing process	Shift manager measures refrigerated product temperatures with a thermocouple or thermometer	Determine maximum length of time held above 40°F (4.4°C); if over 2 hours, discard	Daily inspection audit
					If less than 2 hours, move product to a refrigeration unit capable of maintaining product at 40°F (4.4°C) or lower	
	Cross-contamination		Store covered and away from or below ready-to-eat foods	Observation	Evaluate for signs of contamination	
				Observation	Discard contaminated product	
				Observation	Move product to lower shelf or another refrigerated unit	

EXHIBIT 4.4—continued   **Step: Cooking**

OPERATIONAL STEP	HAZARD	CCP	STANDARD (CRITERIA)	TYPE OF MONITORING	CORRECTIVE ACTION IF STANDARD NOT MET	RECORDS
Cooking ground beef	Bacterial survival due to inadequate cooking	CCP	Cook to 155°F (68.3°C) or higher for at least 15 seconds	Cook will measure end cooked product internal temperature with a thermocouple or thermometer	Continue cooking to 155°F (68.3°C) or higher for at least 15 seconds	Cook's log
	Cross-contamination		Keep raw ground beef away from the nearly or cooked beef on the grill	Observation	Discard contaminated product	
			Wash hands after handling raw ground beef	Observation		
			Use cleaned and sanitized equipment, utensils, and cooking surfaces	Supervision and one-on-one training		
Cooking chili	Bacterial survival due to inadequate cooking	CCP	Cook all ingredients to 165°F (73.9°C) or higher	Cook will measure end cooked product internal temperature with a thermocouple or thermometer	Continue cooking chili to 165°F (73.9°C) or higher for at least 15 seconds	Cook's log
	Contamination from cook's hands or mouth		Use proper tasting procedures	Observation	Discard contaminated product	
				Supervision and one-on-one training		

OPERATIONAL STEP	HAZARD	CCP	STANDARD (CRITERIA)	TYPE OF MONITORING	CORRECTIVE ACTION IF STANDARD NOT MET	RECORDS
					Review proper tasting procedures with cook and check cook's training record	
	Cross-contamination		Use clean and sanitized utensils for stirring	Observation	Wash, rinse, and sanitize all utensils	
				Supervision and one-on-one training	Review proper procedures with cook and check cook's training record	

EXHIBIT 4.4—continued

OPERATIONAL STEP	HAZARD	CCP	STANDARD (CRITERIA)	TYPE OF MONITORING	CORRECTIVE ACTION IF STANDARD NOT MET	RECORDS
Holding chili for service	Bacterial growth	CCP	Hold chili at an internal product temperature of 140°F (60°C) or higher; stir often to maintain even temperature throughout product	Cook measures hot-holding product temperatures with a thermocouple or thermometer	Determine maximum length of time that chili was held below 140°F (60°C); if more than 2 hours, discard If less than 2 hours, reheat chili on range to an internal product temperature of 165°F (73.9°C) or higher for at least 15 seconds	Hot-holding time-temperature log
			Preheat hot-holding unit	Cook measures temperature of water with a thermocouple or thermometer	Continue heating hot-holding unit	
	Cross-contamination		Use cleaned and sanitized equipment and utensils to transfer chili to hot-holding pans	Observation Supervision and one-on-one training	Wash, rinse, and sanitize equipment and utensils following standard operating procedures	
	Bacterial survival and growth		Prepare chili only for same day service	Observation	Review proper procedures with employee, and check employee training record Discard leftover product	

EXHIBIT 4.4—continued

Step: Cooling

OPERATIONAL STEP	HAZARD	CCP	STANDARD (CRITERIA)	TYPE OF MONITORING	CORRECTIVE ACTION IF STANDARD NOT MET	RECORDS
Cooling chili for storage	Bacterial growth Cross-contamination	CCP	Cool chili rapidly from 140° to 40°F (60° to 4.4°C) or lower within 2 hours	Cook will measure internal product temperature with a thermocouple or thermometer every 2 hours during the cooling process	Discard product that does not cool to 40°F (4.4°C) in 4 hours. If product has not cooled to 40°F (4.4°C) or lower within 2 hours, reheat to 165°F (73.9°C) or higher and serve immediately	Cook's cooling log
			Place chili in shallow pans with a product depth of 2 inches or less	Observation	If product depth exceeds 2 inches, move to shallow pans and provide additional pans as necessary so product depth is 2 inches or less	
			Place shallow pans of chili into ice bath. Immerse pans into ice up to product level with the pan and stir frequently	Observation	Prepare ice bath; immerse shallow pans in ice up to the product level within the pan and stir frequently	
	Cross-contamination		After cooling is achieved, cover pans with plastic wrap and put on top shelf of refrigerated unit	Observation	Evaluate for contamination; discard suspect product. Cover and move to upper shelf	
	Bacterial growth during prolonged storage		Label with date, time, and product name	Observation	Label with date, time, and known product name; discard product	

EXHIBIT 4.4—continued

OPERATIONAL STEP	HAZARD	CCP	STANDARD (CRITERIA)	TYPE OF MONITORING	CORRECTIVE ACTION IF STANDARD NOT MET	RECORDS
Reheating chili for service	Bacterial survival and growth	CCP	Rapidly heat chili to an internal product temperature of 165°F (73.9°C) or higher for at least 15 seconds in 2 hours or less	Cook measures internal product temperature with a thermocouple or thermometer	If 165°F (73.9°C) or higher chili temperature for at least 15 seconds is not reached within 2 hours, discard product	Cook's cooking log
Hot-holding chili	Bacterial survival and growth	CCP	Maintain chili temperature at 140°F (60°C) or higher in a preheated hot-holding unit. Stir often to maintain even temperature	Shift manager measures internal hot-holding temperature with a thermocouple or thermometer	If chili temperature is less than 140°F (60°C), discard product	Hot-holding time and temperature log
			Preheat hot-holding unit before chili is placed in the well	Cook measures temperature of water in heat well with a thermocouple	Continue heating hot-holding unit	
			Only reheat chili once	Observation	Discard all leftover products	

HACCP Training

Assess each job and recipe for special training requirements in handling food, operating equipment, and cleaning. Develop specific training objectives for employees based on what they need to do to keep food safe.

Task Analysis

Break down each job into specific duties and HACCP procedures. Add these to your job descriptions. If an employee learns from the start that taking the final cooking temperature of a food is part of the job and not doing this task will hurt a pay raise, promotion, or get him or her fired, he or she is more likely to take the task seriously.

Learning Objectives

The learning objectives should state clearly the key skills an employee needs to correctly do the job, such as learning how to correctly calibrate and use a thermometer and fill out time/temperature logs.

Corrective Actions

Employees need to know what corrective actions they will be allowed to do on their own if a CCP is not being met. Be clear in training so employees know what they may do on their own.

Developing Your Training Program

Your overall approach to training should balance the importance of the topic with the need to make it interesting. Employees learn food safety most easily when they can put what they learn into action right away. They need to see or hear information, practice applying it, and receive quick feedback on how well they are doing.

1. Tell them what they are going to be taught and why it is important to learn.

2. Present the material in various ways, including role-playing and videos.

3. Demonstrate steps and procedures.

4. Answer questions seriously and right away.

5. Let them practice or at least discuss what is presented.

6. Give feedback on their practice performance.

7. Review the material.

8. Test or evaluate performance.

9. Follow up training by monitoring.

10. Retrain if needed.

Choosing Training Methods

There are two basic methods for putting your program into action—individual training and group training. *Individual* or *one-on-one* training assigns one or two trainees to an experienced employee to learn the task. The advantages of this method are that training is focused solely on the job to be done and the trainees receive personal attention, a chance to apply the skill at once, and quick feedback. The disadvantages are the possibility that the trainer is not able to properly tell the trainees how to do the job, the trainees may pick up the trainer's bad habits, and the skill may not be learned.

Group training involves a group of trainees meeting with a trainer, usually in a session apart from their normal work. The advantages are that trainees can work with each other, discuss ideas and together solve problems, and a uniform program can be developed for all trainees. Trainers also can use group exercises, such as games, role-playing, and team competitions, to provide active learning. The disadvantages include trainees get less personal attention and fewer chances to immediately try out new skills. Trainers also may be forced to lecture more than is good for holding the trainees' attention.

Using the best elements from both of these training methods is recommended. One other method, *crash training*, the attempt to cover a lot of material on the job in a very short time, is not recommended for a topic as complex as food safety.

Evaluating Your Training Program

In a HACCP system, on-the-job performance is the key measure of training success. After the training session, pay special attention to the results of your HACCP monitoring procedures. Ask yourself:

1. Did the training produce results on the job? Are employees meeting the training objectives related to CCPs and standards?

2. If the intended results were not produced, why not?

If you are not getting the results you intended, you may decide to work with the employees individually, ask for their opinion as to why it is not working, schedule more training, or take another course of action.

A Case in Point

At 7:00 AM, Jason, an employee at Cal's Catering, came in to work. Three hours later he was done preparing 50 roast beef sandwiches that were to be delivered to a nearby office at noon. Placing the sandwiches in a large, clean delivery box, Jason set them on the counter in the kitchen at a room temperature of 85°F (29.4°C). After removing his plastic gloves and washing his hands, Jason began preparing vegetable trays. Once those were done and refrigerated, Jason prepared the fruit trays. Shortly before noon, Jason finished them. He called the delivery driver to take the food to the office party.

The food was greatly enjoyed and the party was a rousing success—until early the next morning, when some partygoers called in sick.

At Cal's Catering what food safety factor was broken?

How might Cal's Catering adjust its procedures to make sure that this error is not repeated?

Answer to a Case in Point

Roast beef, a potentially hazardous food, must be kept out of temperatures that support the growth of bacteria. Jason kept the beef unrefrigerated for at least five hours as he prepared sandwiches and the other food. Although he did refrigerate the vegetable trays that he prepared, he failed to refrigerate the sandwiches, which remained in the kitchen at a room temperature of 85°F (29.4°C).

To make sure that this error is not repeated, Jason's manager needs to let his or her employees know what the time limit is for beef to remain unrefrigerated during preparation. Procedures that require the refrigeration of sandwiches and specify the number to be prepared at one time need to be written into the recipe. Jason's manager can then actively involve Jason and his co-workers in monitoring the temperature of the beef.

Chapter 4 Exercise

1. In a HACCP system, what is the key measure of training success?

    a. Employee obedience.

    b. Price per entree.

    c. Number of employees to number of managers.

    d. On-the-job performance in food safety.

2. When you first set up a HACCP system, introduce it to employees by:

    a. telling them that those not adapting to the system will be moved to nonfood prep duties.

    b. assigning them to memorize all HACCP terms.

    c. keeping the lines of communication open, so employees understand what they must do and feel free to ask questions.

    d. putting a notice on the bulletin board.

3. Food safety training objectives for foodhandlers should be centered on:

   a. CCPs and standards.
   b. decreasing food waste.
   c. more expensive entrees.
   d. controlling portion size.

4. Trainees learn best when they:

   a. are given written material.
   b. hear a lecture.
   c. practice applying it.
   d. use trial and error on the job.

5. HACCP food safety systems focus on:

   a. protecting food at all times during the flow of food.
   b. verifying that produce is fresh when delivered.
   c. protecting meat from mold.
   d. producing food only in large quantities.

6. Flowcharts are:

   a. recipes with more than seven ingredients.
   b. time and temperature logs.
   c. diagrams that show the flow of food.
   d. charts showing the growth of bacteria.

7. Which one of the following should you do if you find a recipe is too difficult to handle as it is written?

   a. Make sure you are the one to prepare it.
   b. Simplify the steps or ingredients in it.
   c. Offer the menu item for one week only.
   d. Lower the price on the menu.

8. The practice of entering a temperature in a log without really measuring the temperature is called:

   a. monitoring.
   b. verifying.
   c. standard setting.
   d. dry lab.

9. Which one of the following is a *corrective action* for handling chili?

   a. Hold chili at 140°F (60°C) or higher until served.
   b. Cool chili from 140° to 40°F (60° to 4.4°C) or lower within four hours.
   c. Reheat chili to an internal temperature of 165°F (73.9°C) or higher if held less than two hours below 140°F (60°C).
   d. Place chili in shallow pans with a product depth of 2 inches or less into an ice bath.

10. Temperature logs should be kept:

   a. in your office.
   b. near employee work areas.
   c. in one central location.
   d. near employee break areas.

CHAPTER 5: ADAPTING HACCP PRINCIPLES FOR YOUR OPERATION

Test Your Food Safety IQ

1. **True or False:** Very young, elderly, and ill diners may be especially vulnerable to foodborne illnesses. (See *Institutional Service Operations*, page 62.)

2. **True or False:** Provide customers with short-handled serving spoons at a food bar. (See *Food Bars and Other Self-Service Arrangements*, page 63.)

3. **True or False:** It is acceptable to reuse cardboard boxes as delivery containers for food. (See *Off-Site Delivery*, page 65.)

4. **True or False:** Caterers must meet the same food safety standards as permanent operations. (See *Catering*, page 67.)

5. **True or False:** Vending machines can be installed anywhere in a foodservice operation. (See *Vending Machines*, page 68.)

Learning Objectives

After completing this chapter, you should be able to:

◆ Understand the basic food safety needs of quick service, full service, and institutional service operations.

◆ Adapt HACCP principles to the types of service your operation provides.

Quick Service Operations

Quick service operations usually feature short waiting times for service, limited menus, and service counters where customers wait for, pick-up, and pay for their food. Such operations range from small stands to large sites with drive-through and sit-down dining facilities. Full service and institutional service operations may offer carryout service, which is similar to quick service.

Standardize procedures:

◆ Focus training on foodhandling, cooking times and temperatures, and personal hygiene.

◆ Cook all potentially hazardous foods to safe internal temperatures.

◆ Prepare only small batches of food in advance.

◆ Regulate fat content, size, and thickness of each portion to predict cooking time.

- ◆ Regularly check gas jets and preheat grills for uniform temperatures.
- ◆ Consider pre-packaged salads as an alternative to self-service salad bars.
- ◆ Separate foods when packaging them to prevent temperature changes and cross-contamination. Separately package condiments and plastic utensils.

Full Service Operations

Full service operations may have extended menus and many items may require multi-step preparation. In this chapter, see *Food Bars and Other Self-Service Arrangements*. See *Chapter 9* for information on employee practices and on safeguarding food, customers, and table settings.

Institutional Service Operations

Institutional service operations include nursing homes and hospitals, childcare facilities and schools at all levels, and corporate dining rooms and cafeterias. These operations usually serve groups of customers and may need to provide special diets and off-hour meals. Managers often plan menus and order supplies well in advance.

Nursing homes, hospitals, and schools must be aware that they serve customers—the young, elderly, and ill—who may be especially vulnerable to foodborne illnesses.

1. Food Choice. Use only:

   - ◆ Government-approved, commercially processed foods. *(Never use homemade and home-canned foods.)*

   - ◆ Pasteurized milk and milk products.

   - ◆ Pasteurized eggs—to guard against *Salmonella*. If you use shell eggs, cook them until they are firm.

   - ◆ Accepted amounts of food additives and preservatives. (See *Additives and Preservatives* in *Chapter 2*.)

2. On-Site Delivery. To keep foods safe while in transit:

   - ◆ Sanitize trays and utensils. Store them in clean areas or sanitized containers.

   - ◆ Separate foods when packaging them to prevent temperature changes and cross-contamination. Separately package condiments and plastic utensils.

   - ◆ Use containers designed to maintain temperatures when transporting food.

- Use delivery routes that quickly move food to assembly kitchens or pantries without passing through extremely hot or cold areas.

- Use delivery equipment, such as delivery carts, that have been cleaned and sanitized.

- Reheat hot foods to 165°F (73.9°C) for at least 15 seconds, then hold at 140°F (60°C) or higher. Deliver cold foods at 40°F (4.4°C) or lower.

Food Bars and Other Self-Service Arrangements

Food bars, buffets, and cafeterias usually include a display of hot or cold foods. Customers walk by, either help themselves or are served by attendants, and then bring their food back to their tables. Customers often return for more servings. Food often is displayed for extended time periods in high-traffic areas. Special precautions are needed to control temperatures and physical contamination.

Control display procedures, amount served, and customer behavior:

- Label all items so that customers do not need to sample or return them.

- Reheat hot foods to 165°F (73.9°C) for at least 15 seconds, then hold at 140°F (60°C) or higher. Hold cold items at 40°F (4.4°C) or lower (see *Exhibit 5.1*).

- Put ready-to-eat displayed foods on plates—not directly on ice. Ice surrounding chilled items must drain away from food containers. Sanitize drip pans after each use.

- Check food temperatures with a thermometer or thermocouple every two hours. Record temperatures in a log.

- Near chilled items, use lighting that will not raise food temperatures. Place plastic shields around lights to guard against broken glass or use plastic-coated lights.

- Provide tongs or a long-handled ladle for each item so customers do not touch the food.

- Place plastic sneeze guards or shields—at a minimum of 14 inches but no higher than 48 inches above the food—in a direct line between the customer's face and the food.

- Do not reuse ice, vegetable, or plant decorations that have been soiled by food.

- Servers should:

  - Give customers clean tableware each time they go through the food line.

  - Alert customers about supervising their children.

- Remove any food containers, serving dishes, and utensils that customers have touched, tasted, or possibly contaminated in another way.

EXHIBIT 5.1 HOT AND COLD HOLDING UNITS

Hot Holding Unit

Cold Holding Unit

Outdoor Service

Outdoor service includes table service, food bars, and other self-service operations. Check with your local or state health department for specific regulations on outdoor service, overhead protection, and enclosure. Supervise food choice, production, temperatures, packaging, and service:

◆ Hold cold foods at 40°F (4.4°C) or lower. Check food temperatures with a thermometer or thermocouple every two hours, more often in hot weather. Record temperatures in a log.

◆ Hold hot foods at 140°F (60°C) or higher. Check food temperatures with a thermometer or thermocouple every two hours. Record temperatures in a log.

◆ For outdoor buffets, prepare smaller batches of each item and whether or not used up, regularly discard and replace those batches. Do not mix new food into old.

◆ Put ready-to-eat displayed foods on plates—not directly on ice. Ice surrounding chilled items must drain away from food containers. Sanitize drip pans after each use.

◆ In hot weather, check ice-making equipment and arrange for enough staff to provide fast service.

◆ Use chilled plates to serve cold items.

- Serve condiments, such as ketchup, in sealed containers or only on request.

- Provide wind screens to keep dirt and pests (see *Chapter 11*) out of food.

- Set up foodservice and dining areas away from portable toilets and privies. At fair grounds, set up away from manure piles.

Central Kitchens

Central kitchens are often used to prepare food for large groups. They may serve food on-site or provide for services such as off-site delivery, catering, mobile and temporary units, and vending machines. (These operations are discussed later in this chapter.)

Central kitchens must use the proper procedures and equipment for chilling and holding. (In this chapter, see *Institutional Service* and *Off-Site Delivery*.)

- Provide:

  - Equipment for deep-chilling large quantities of food at 26° to 32°F (–3.3° to 0°C).

  - Refrigerators for short-term storage at a food product internal temperature of 40°F (4.4°C).

  - Freezers for storing already chilled or frozen foods at 0°F (–17.8°C) or lower.

- Operations that need to cook meat, such as rare prime rib, lower than 140°F (60°C) for 12 minutes or 130°F (54.4°C) for 121 minutes must use special equipment.

Off-Site Delivery

Since foods are often cooked but not immediately consumed many possibilities exist for food to be contaminated and time-temperature abused during delivery. Control production, holding, packaging, and delivery:

- Cook hot foods to a proper internal product temperature, then hold at 140°F (60°C) or higher.

- Deliver cold items at 40°F (4.4°C) or lower.

- Package food in small batches to prevent temperature changes and cross-contamination. Separately package condiments and plastic utensils.

◆ Covered food containers must sustain food temperatures, allow air to circulate, keep food from spilling or leaking, and be disposable or easy to clean and stackable. Only use containers designed to transport food—do not reuse cardboard boxes as food containers (see *Exhibit 5.2*).

**Consider packing an extra meal and measuring its temperature at the end of the delivery route to determine how well your equipment protects the food.

◆ Plan routes so food is delivered within a safe time and at a safe temperature.

◆ Quickly load foods.

◆ Label foods with proper storage, shelf life, and reheating instructions for employees at off-site locations and customers.

◆ Keep delivery vehicles clean and well-maintained. (Breakdowns may spoil food.) Servicing areas for delivery vehicles, such as garages, should be clean, dry, and away from food storage and preparation areas.

EXHIBIT 5.2 PROPER CONTAINERS FOR TRANSPORTING FOOD

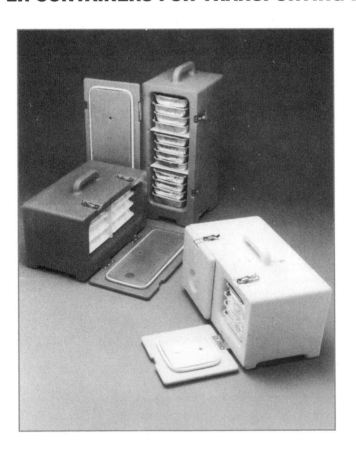

Courtesy of Cambro Manufacturing Company, Huntington Beach, California.

Catering

Caterers provide food for airlines, private parties and events, and public and corporate affairs. Caterers may bring in ready-to-eat food or may prepare food in a mobile or temporary unit, in rented facilities, or with the customer's own equipment. Review the guidelines discussed in *Central Kitchens* and *Off-Site Delivery*. In addition:

◆ Reheat hot foods to 165°F (73.9°C) for at least 15 seconds within two hours, before holding and serving at 140°F (60°C).

◆ Check all foods with a thermometer or thermocouple every two hours. Record temperatures in a log.

◆ Be sure there is a safe drinkable water supply (including enough hot water) for cooking and washing and adequate toilet and handwashing facilities. (See *Chapter 3*.)

◆ Be sure there is enough power to run your cooking and cooling equipment.

◆ Be sure there are adequate facilities for garbage disposal.

◆ Check for signs of insects and rodents. Refuse to prepare or serve food in an area that cannot be protected against pests.

◆ For outdoor catering, such as barbecues and cookouts:

• Deliver raw meat frozen, wrapped, and on ice. Consider supplying ice chests if customers themselves wish to transport food. Chill these containers before filling them with potentially hazardous foods.

• Deliver all milk products in a refrigerated vehicle or on ice.

• Take special care when handling potentially hazardous foods, such as meat or fish salads. If used, pack them on ice or in a refrigerated device, separate from bread and fillings, such as lettuce.

• *Never* allow homemade or home-canned foods to be served.

• Supervise customers' behavior around cooking equipment and food displays.

• If food is left with the customer after the catered engagement, provide proper storage, shelf life, and reheating instructions.

Mobile Units

These include driveable and portable serving and preparation facilities. They may range from soft-drink stands to elaborate field kitchens. If the unit serves only beverages and ready-to-eat packaged foods, health requirements may be relatively simple. If the unit prepares and serves potentially hazardous foods, all HACCP practices necessary for permanent foodservice operations are required.

Temporary Units

The FDA defines a *temporary unit* as a unit that is licensed to operate in a specific location for a certain period of time, usually not exceeding 14 consecutive days, in connection with a single event or celebration. Licenses may extend for an entire season. As with mobile units, local health departments usually specify the requirements these units must meet. In most cases, temporary units should not cook potentially hazardous foods. Exceptions include pre-prepared or pre-packaged foods and foods that require limited preparation, such as hot dogs. If local laws allow the preparation of potentially hazardous foods, review the requirements for *Catering and Off-Site Delivery*. In addition:

◆ Prepare food in a commercial kitchen under sanitary conditions.

◆ Transport foods to the temporary kitchen, store them, and keep them out of the temperature danger zone. No thawed, ready-to-eat potentially hazardous foods should be delivered or stored in direct contact with water or ice.

◆ Pre-package food for individual service.

◆ Have drinkable water available for cleaning, sanitizing, and handwashing. If utensils cannot be sanitized, single-service articles must be used.

◆ Enclose service counter areas using tight-fitting solid or screened doors. Service windows should be open only when in use. If this is not possible, use fans or limited service, moving food from the grill directly to the bun.

◆ The unit must protect food and equipment from weather and contamination.

Vending Machines

Vending machines are money-, card-, or key-operated self-service devices that store and dispense food or beverages. These machines can carry a variety of hot and cold foods, including potentially hazardous foods. These foods are often ready-to-eat and packaged from a supplier. If you have vending machines, you are responsible for protecting the foods that are dispensed. Choose the proper facilities and equipment protection:

◆ Carefully choose vending suppliers. (See *General Purchasing Guidelines* in *Chapter 6*.)

◆ Check that vending supplies are properly packaged by your supplier:

- Fruits and vegetables that have an edible peel or outer surface, such as apples or celery, must be washed, dried, and wrapped before being placed in the vending machine.

- Foods must be stored in sealed, moisture-resistant packages. Packaged potentially hazardous foods (such as wrapped sandwiches to be microwaved) must be dispensed in their original containers or wrappers.

◆ Select machines that:

- Keep food out of the temperature danger zone. Each machine with hot or cold storage capacity must have an automatic system that shuts down the vending mechanism if the temperature is out of the safety zone.

- Have food-contact surfaces that are easily cleanable, corrosion-resistant, and non-absorbent.

- Are National Automatic Merchandising Association (NAMA) listed (or the equivalent).

◆ Train employees to use good personal hygiene when servicing and refilling machines. Employees should wash their hands before and after these tasks.

◆ Install machines away from garbage containers, sewer drains, and pipes and in areas where ceilings, walls, and floors can be kept clean and pest-free.

◆ Be sure there is a safe supply of drinkable water for beverage machines. Guard against cross-connections from faulty or corroded pipes (see *Chapter 9*).

◆ Secure an adequate and dependable power supply for cooking and cooling equipment.

A Case in Point

On February 2, 1975, one crew member and the 196 passengers aboard a chartered plane flying from Tokyo to Copenhagen, with an interim stop in Anchorage, developed an illness characterized by diarrhea, vomiting, abdominal cramps, and nausea.

Breakfast was served $5\frac{1}{2}$ hours after the plane left Anchorage. The meal consisted of an omelet, ham slices, yogurt, bread, butter, and cheese. The passengers began getting ill approximately two hours after the meal was served.

Except for the crew member who had sampled the ham from the passengers' breakfast, none of the plane's crew became ill. Instead of breakfast, the crew had been served a steak dinner since it was supper time for them.

The breakfast had been prepared in Anchorage by a catering company. Three cooks were involved in the preparation of the ham and omelets. The ham had been prepared the night before and then refrigerated. The next day, one of the cooks placed the ham slices on top of the omelettes as they were being prepared. The food was kept at room temperature during the six hours spent preparing the omelettes.

Following preparation, this food was in a holding room for 14$^1/_2$ hours at 50°F (10°C). Beginning at about 7:30 AM, the breakfast food was loaded onto the plane. The food was again stored at room temperature until it was heated just before serving.

Ham that had been handled by a cook who had an inflamed cut on his finger caused this outbreak. *Staphylococcus aureus* was found in the cook's cut.

What else went wrong that contributed to the outbreak?

Answer to a Case in Point

The ham was held at room temperature long enough to allow the staphylococcal bacteria to multiply and produce toxins. Spending 20 hours at room temperature made the ham the source of one of the largest foodborne outbreaks ever recorded.

Staphylococcal toxins can not be destroyed by ordinary cooking temperatures. The best control measure is to prevent this bacteria from entering the food at all. This case emphasizes the risk of storing food at room temperature for long time periods, as well as the risk of allowing infected foodhandlers to prepare food.

Chapter 5 Exercise

1. When holding hot food, quick service and carryout operations should:

   a. limit holding time and regularly check temperatures.

   b. hold large batches at temperatures as low as possible.

   c. hold hot foods at 70°F (21.1°C).

   d. use holding equipment to reheat food.

2. If an institution serves eggs to very young, elderly, or ill diners, it should:

   a. store the eggs at room temperature.

   b. use pasteurized eggs to guard against salmonella.

   c. only use shell eggs.

   d. offer lightly cooked shell eggs in omelettes and scrambled eggs.

3. At a self-service food bar, ready-to-eat foods should be placed:

   a. directly on ice.

   b. near high-intensity lights.

   c. on sanitized chilled plates on top of ice.

   d. outside of the sneeze guard.

4. When serving food outdoors:

   a. mix cold food into warm food.

   b. serve ketchup and mayonnaise in large, uncovered bowls.

   c. put ready-to-eat food directly on ice.

   d. serve cold items on chilled plates.

5. Food to be delivered hot should be transferred to delivery containers:

   a. in the early morning for afternoon deliveries.

   b. at proper internal product temperatures.

   c. at 40°F (4.4°C).

   d. in reused cardboard containers.

6. During a catering job, if your customer asks you to serve home-canned tomatoes:

   a. accept the tomatoes and heat them to 140°F (60°C).

   b. accept the tomatoes and cool them to 40°F (4.4°C).

   c. reject the tomatoes unless they are canned in glass.

   d. reject the tomatoes because home-canned foods must never be used.

7. During a catering job, if you find signs of rodents in the kitchen area:

   a. refuse to prepare or serve food in that area.

   b. have your employees wear gloves.

   c. immediately set traps to catch them.

   d. cover all holding pans.

8. In a temporary kitchen, if knives, forks, and spoons cannot be sanitized:

   a. wipe them on a clean towel before use.

   b. rinse them in warm water.

   c. use single-service utensils.

   d. spray mild disinfectant on them, then wipe them off.

9. Vending machines with hot or cold storage must have:

   a. no plastic parts.

   b. an automatic system that shuts down the vending mechanism if the temperature is out of the safety zone.

   c. a supply of plastic gloves.

   d. no display windows for food.

10. In a vending machine, packaged potentially hazardous foods (such as wrapped sandwiches to be microwaved) must be:

    a. stored at 70°F (21.1°C).

    b. removed from their wrapper before being put in the machine.

    c. dispensed in their original containers or wrappers.

    d. wrapped in metal foil.

CHAPTER 6: PURCHASING AND RECEIVING SAFE FOOD

Test Your Food Safety IQ

1. **True or False:** Look for suppliers who use HACCP systems. (See *General Purchasing Guidelines*, page 72.)

2. **True or False:** To save time, schedule all deliveries for the busiest time of the day. (See *Receiving*, page 81.)

3. **True or False:** Glass thermometers can be substitutes for metal thermometers. (See *Types of Thermometers*, page 76.)

4. **True or False:** Game meats must only be purchased from suppliers and producers who follow governmental regulations. (See *Meats and Game Meats*, page 73.)

5. **True or False:** Buy shellfish only from government-approved suppliers. (See *Fish and Shellfish*, page 74.)

Learning Objectives

After completing this chapter, you should be able to:

◆ Set up purchasing and receiving standards and procedures.

◆ Choose reliable suppliers.

◆ Provide receiving facilities and equipment.

◆ Use food thermometers.

◆ Purchase and inspect specific foods.

◆ Reject shipments.

General Purchasing Guidelines

Safely start the flow of food by choosing reliable suppliers who meet federal and local standards. Visit suppliers and inspect their facilities and procedures. Then work closely with your suppliers to set up proper procedures.

Meats and Game Meats

Food services are required to buy only meat that has been inspected by the United States Department of Agriculture (USDA) or by a state department of health (see *Exhibit 6.1*). Ask your supplier for written proof of government-inspected meat.

Game meats include meats not classified as cattle, sheep, swine, or goat. Examples are deer, rabbit, wild ducks, and unusual meats, such as alligator and snake. These meats may have been commercially raised or killed in the wild. In all cases, they must only be purchased from suppliers and producers who have followed governmental regulations concerning raising, slaughtering, inspecting, and processing.

The USDA can also grade meats. This stamp identifies the relative quality of the meat.

EXHIBIT 6.1 MEAT STAMPS

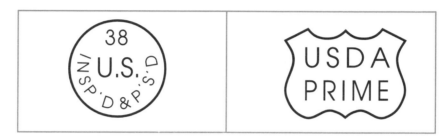

Poultry

Only purchase poultry that carries a federal or state inspection stamp on the carcass, parts, or packaging (see *Exhibit 6.2*). A grade stamp indicating quality also may appear on the carcass or packaging.

EXHIBIT 6.2 POULTRY INSPECTION STAMPS

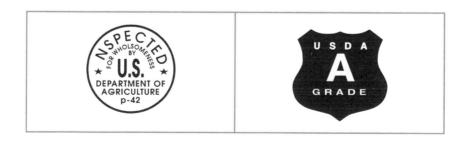

Fish and Shellfish

Purchase all fish only from suppliers that are approved by regulatory authorities. Suppliers may have to obtain a regulatory variance to harvest, perpare, and distribute fish that may cause scombroid intoxication or ciguatera. (See *Chapter 2.*)

Shellfish must be bought only from suppliers on Public Health Service FDA lists of Certified Shellfish Shippers or on lists of state-approved sources. Shellstock identification tags must be kept on file for 90 days after clams, mussels, and oysters have been received (see *Exhibit 6.3*).

Eggs

Buy eggs only from approved suppliers. Purchase only a one- or two-week supply. Your purchase order should require:

◆ Government-inspected Grade AA or A eggs with the USDA shield on the carton. All pasteurized egg products—liquid, frozen, and dried—must have an inspection stamp. (See *Exhibit 6.4*.)

◆ Refrigerated storage and delivery.

◆ Eggs that are less than two weeks old. All shipments must carry use-by dates.

Dairy Products

Buy only pasteurized milk. All milk products, such as cream, dried milk, cottage cheese, Neufchatel cheese, and cream cheese, must also be made from pasteurized milk.

Modified Atmosphere (MAP) and Sous Vide Packaged Foods

These types of sealed packaging reduce or replace the oxygen with other gases. Refrigerated entrees, fresh pasta, prepared salads, soups, sauces and gravies, cooked or partially cured meats, and poultry dishes are available packaged like this. These types of packaging allow longer shelf life and foods are kept fresh and wholesome by their packaging and refrigeration rather than from preservatives and additives.

Although spoilage is almost entirely prevented, if the product is mishandled the inside atmosphere may promote the growth of bacteria that can cause botulism or listeriosis. For these reasons, purchasing is a *critical control point* (CCP) for MAP or sous vide products.

EXHIBIT 6.3 SHELL-STOCK IDENTIFICATION TAGS

THIS TAG IS REQUIRED TO BE KEPT ON CONTAINER UNTIL EMPTY AND THERE-AFTER KEPT ON FILE FOR 90 DAYS

TO BE RETAINED BY RECEIVER FOR 90 DAYS

THIS PACKAGE CONSISTS OF OYSTERS

Gals   Bu

Packed by

Address: **Bon Secour, Ala.**

Distributed by _____

Address _____

Date Reshipped _____

SHELLFISH DREDGED FROM

BAY GARDENE

LOCAL AREA OR BED NO.   **LA 51**

DATE

01/10/84

TO:

OYSTERS

ALABAMA STATE BOARD OF HEALTH

BUREAU OF SANITATION

DIVISION OF INSPECTION
MONTGOMERY, ALA.

No. 00576

SHIPPER'S NAME AND ADDRESS

FROM

Bon Secour, AL 36511

Certificate No.   Ala 49

BELOW TO BE FILLED IN BY RECEIVER

DATE

REC'D

LOT NO	LOT CONSISTS OF

Courtesy of the Alabama Department of Public Health, Seafood Branch, Mobile, Alabama

EXHIBIT 6.4 USDA EGG SHIELD

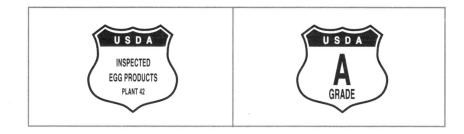

Fresh Produce

Some plants, such as wild mushrooms, may carry harmful toxins. Purchase them only from suppliers who can prove the plants were individually inspected. In some areas, pre-washed salad items, such as lettuce, may be available from dependable vendors. Tell the supplier when and how items are to be prepared and delivered.

Other General Purchasing Procedures

Look for suppliers who:

◆ Use properly refrigerated delivery trucks.

◆ Train their employees in sanitation.

◆ Use protective, leak-proof, sturdy packaging.

◆ Adjust delivery schedules so goods do not arrive during busy periods.

◆ Cooperate with your employees in inspecting food when it is delivered.

◆ Allow you to inspect their delivery trucks and production facilities.

Types of Thermometers

Temperature control is the single most important aspect of food safety. Your employees need to be able to use and take care of several types of food thermometers. (See *Chapter 7* for information on storage area thermometers.)

Use metal food thermometers. (Mercury-filled or glass thermometers may break.) Food thermometers should be able to measure internal temperatures from 0° to 220°F (−17.8° to 104.4°C). They should be accurate to ±2°F or ±1°C. Common types of food thermometers include the following:

Thermocouples

Thermocouples measure temperature through a sensor in the tip of the stem (see *Exhibit 6.5*). When the user presses a button, the thermocouple produces a readout of the temperature. This device accurately and quickly measures a range of temperatures without the need to recalibrate (adjust for accuracy) often.

EXHIBIT 6.5 THERMOCOUPLE

Bi-metallic Stemmed Thermometers

Bi-metallic stemmed thermometers are the most common type of foodservice thermometers (see *Exhibit 6.6*). They measure temperature through a metal stem with a sensor in the lower end. The sensing area is from the tip to a half-inch past the dimple. When selecting and using this type of thermometer, remember that it should have:

◆ An adjustable calibration nut.

◆ Easy-to-read numerical temperature markings.

◆ A dimple marking the end of the sensing area (which begins at the tip).

Digital Thermometers

Digital thermometers measure temperatures through a metal tip or sensing area and provide a digital readout (see *Exhibit 6.7*). They are especially easy to read. Many models are available for measuring surface, interior, and air temperatures.

Time-Temperature Indicators (TTIs)

Liquid crystals in strips that change color when packaged contents reach an unsafe temperature are called *time-temperature indicators* (TTIs) (see Exhibit 6.8). They are often used with sous vide, MAP, or cook-chill packaging. (See *Modified Atmosphere* [MAP] and *Sous Vide Packaged Foods* in this chapter.)

EXHIBIT 6.6 BI-METALLIC STEMMED THERMOMETER

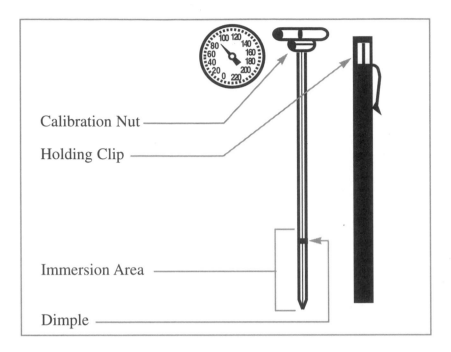

Calibration Nut

Holding Clip

Immersion Area

Dimple

EXHIBIT 6.7 DIGITAL THERMOMETER

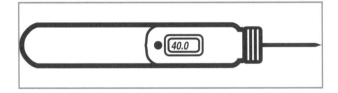

EXHIBIT 6.8 TIME-TEMPERATURE INDICATOR (TTI)

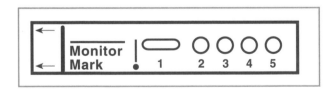

Other Food Thermometers

Meat and deep-fry thermometers measure temperatures in specific types of food. Candy thermometers are usually accurate to ±5°F or ±2.8°C and are not accurate enough to be used to measure other foods.

Calibrating Thermometers

You need to make sure your thermometer readings are accurate (see Exhibit 6.9). Recalibrate thermometers regularly, after an extreme temperature change, or if the unit has been dropped. Thermometers may be calibrated by one of two methods—ice point method for cold foods, or boiling point method for hot foods.

Using the *ice point method* you submerge the sensor in a 50/50 ice and water slush. For a bi-metallic stemmed thermometer, wait until the needle stops, then use a small wrench to turn the calibration nut until the thermometer reads 32°F (0°C). For a thermocouple or digital thermometer, try a new battery or have the manufacturer or a repair service check the unit.

Using the *boiling point method* you submerge the sensor into boiling water. For a bi-metallic stemmed thermometer, wait until the needle stops, then use a small wrench to turn the calibration nut until the thermometer reads 212°F (100°C). Follow the same instructions for the thermocouple and digital thermometer that were used with the ice point method. You need to be very careful when using the boiling point method to avoid burns.

Note: The boiling point lowers about 1°F (0.6°C) for each 550 feet above sea level, so you need to adjust your thermometer accordingly.

EXHIBIT 6.9 CALIBRATING THERMOMETERS

Using Food Thermometers

Use the following general procedures:

◆ Wash, rinse, sanitize, and air-dry thermometers before and after each use. A sanitizing mixture or fabric wipe for food-contact surfaces can be used.

- Do not let the sensing area touch the bottom or sides of food containers.

- Insert the stem so that the sensing area is in the center of the food. Wait at least 15 seconds for the reading to steady and then record the reading (see Exhibit 6.10).

- Use the unit to measure frozen, refrigerated, tepid, and hot foods and liquids. *Never* leave the thermometer in food that is being cooked by oven, microwave, or stove.

Receiving

Provide the necessary facilities and equipment. Train your employees to be sure that goods arrive in sanitary conditions and are handled properly. Finally, all food should be moved to storage right away.

Receiving area employees should be trained to:

- Inspect supplies right away.

- Check delivery trucks for any signs of contamination, such as melted ice and dirt.

- Identify the required government inspection stamps.

- Check expiration dates and use-by dates.

- Use thermometers to measure product temperatures. (See *Using Food Thermometers*.)

- Sample all bulk items and individual packages within cases.

- Log in acceptable goods.

- Remove staples, nails, and other fasteners before unpacking boxes and crates.

- Quickly move items to storage. Do not leave them on the dock or in hallways.

- Reject unacceptable goods or contact a supervisor to do so. (See *Rejecting Shipments*.)

It is important to set delivery times so goods do not arrive during busy periods and so several shipments do not arrive at the same time. Make sure only authorized employees sign for deliveries.

Facilities and Equipment

When setting up your receiving area, be sure that it is:

- Clean, well-lit, pest-free, and supplied with copies of receiving rules.

◆ Equipped with sanitized carts, dollies, hand trucks, and containers and has areas for washing, drying, wrapping, and re-wrapping supplies.

EXHIBIT 6.10 TAKING FOOD TEMPERATURES

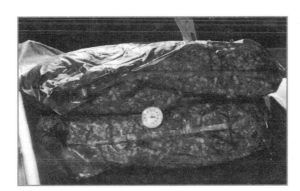
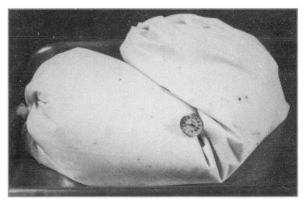

General Receiving Guidelines

Your food safety system should include standards for receiving each kind of food (see *Exhibit 6.11*).

Receiving Special Foods

Shell Eggs

- ◆ Shells must be clean and uncracked.

- ◆ Measure the temperature of a shipment by breaking one or two eggs into a glass and submerging the entire sensing area of the thermometer into the glass. Eggs should arrive at a temperature of 40°F (4.4°C) or lower.

- ◆ Eggs should not give off an odor. Yolks should be high and firm. Whites should cling to the yolk.

- ◆ Immediately refrigerate acceptable eggs in their original containers.

Other Types of Eggs

Frozen eggs should be received at 0°F (−17.8°C). Liquid eggs should be received at 40°F (4.4°C) or lower. Check use-by dates. All packages should be intact.

Dairy Products

All non-frozen dairy products must be delivered at 40°F (4.4°C) or lower. Ice cream may be received at 6° to 10°F (−14.4° to −12.2°C). Dairy products should not have a sour or moldy taste, visible mold or specks, or a strange color or texture. Products should also be delivered so you can use them before the expiration date.

- ◆ *Milk:* Measure the temperature of milk delivered in cases by opening a carton and submerging the entire sensing area of your sanitized thermometer. Discard the sampled carton or use it right away. If the milk is packaged in a bulk plastic container, fold the soft plastic around the thermometer being careful not to poke a hole in the package.

- ◆ *Butter:* Should have a fresh flavor, even color, and firm texture.

- ◆ *Cheese:* Should have the right flavor and texture for the kind of cheese. If the cheese has a rind, it should be clean and unbroken.

EXHIBIT 6.11 ACCEPTABLE AND UNACCEPTABLE CONDITIONS FOR RECEIVING MEAT, POULTRY, FISH, AND SEAFOOD

FOOD	ACCEPT	REJECT
FRESH MEAT (such as beef, lamb, pork) 	**TEMPERATURE:** at 40°F (4.4°C) or lower **BEEF COLOR:** bright, cherry red **LAMB COLOR:** light red **PORK COLOR:** white fat; pink lean portion **TEXTURE:** Firm and springs back when touched	**COLOR:** brown or greenish; brown, green, or purple blotches; black, white, or green spots **TEXTURE:** slimy, sticky, or dry Cartons are broken; meat wrappers are dirty; packaging is torn
FRESH POULTRY 	**TEMPERATURE:** at 40°F (4.4°C) or lower **COLOR:** no discolorations **TEXTURE:** firm and springs back when touched . Should be surrounded by crushed, self-draining ice	**COLOR:** purplish or greenish or green discoloration around the neck; darkened wing tips **ODOR:** abnormal odor **TEXTURE:** stickiness under wings and around joints; soft, flabby flesh
FRESH FISH 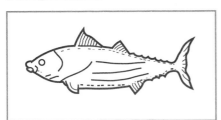	**TEMPERATURE:** at 40°F (4.4°C) or lower **ODOR:** no fishy odor **EYES:** bright, clear, and full **TEXTURE:** flesh and belly are firm and spring back when touched Packed in self-draining ice	**COLOR:** gray or gray-green gills **ODOR:** fishy or ammonia odor **EYES:** sunken, cloudy, or red-bordered **TEXTURE:** dry gills; flesh is soft and gives; if a finger is pressed on the flesh, the fingerprint will stay

EXHIBIT 6.11—continued

FOOD	ACCEPT	REJECT
FRESH SHELLFISH (such as clams, mussels, oysters) 	**TEMPERATURE:** at 45°F (7.2°C) or lower for live shellfish; 0°F (–17.8°C) for frozen products No strong odor **SHELLS:** closed **SHIPPED:** alive Identified by a shell stock tag. Lots should not be mixed up. Record the delivery information, including the dates on the tags. Tags must be kept for 90 days.	Shells that are partly open and do not close when tapped mean the clams, mussels, and oysters are dead.
FRESH CRUSTACEA (such as lobsters, shrimp) 	**TEMPERATURE:** at 45°F (7.2°C) or lower for live lobsters; 0°F (–17.8°C) for frozen products **SHIPPED:** alive No strong odor Lobster shell: hard and heavy	**SHELL:** soft **ODOR:** strong

Modified Atmosphere (MAP) and Sous Vide Packaged Foods

Receiving is a CCP for processed packaged foods (see *Exhibit 6.10*). Reject all packages that do not meet the following standards:

◆ Packages should not have any holes or tears, bubbles, slime, or discolored contents.

◆ Frozen packages should arrive at 0°F (–17.8°C) or lower.

◆ Measure package temperatures by holding the thermometer tightly between two packages, being careful not to poke a hole in them. These foods must be received at the temperature on the package specified by the manufacturer or supplier.

◆ Check any TTI on the package to see if it has changed color.

◆ Check the expiration date or use-by date on the package.

MAP fish must be rejected if there is slime, bubbles, or too much liquid. These may be signs of chemical or bacterial contamination.

Aseptically and Ultra-Pasteurized Packaged Foods

Ultra-pasteurized foods include milk products and fruit juices in cartons that have been heat-treated to kill disease-causing micro-organisms. Foods labeled *UHT* have been *ultra-pasteurized* (high temperature/short time), then *aseptically packaged* (hermetically sealed).

◆ UHT foods may be received unrefrigerated.

◆ UHT milk products must be refrigerated at 40°F (4.4°C) or lower after opening.

Frozen Foods

Carefully check incoming frozen foods:

◆ Frozen foods should arrive in air-tight, moisture-proof wrappings at 0°F (–17.8°C). Ice cream may be received and stored at 6° to 10°F (–14.4° to –12.2°C).

◆ Check the temperature of frozen foods by opening one case and inserting the entire sensing area of the thermometer between two packages, being careful not to poke a hole in them. Reseal, date, and initial the case so other employees know it was opened and checked.

◆ Thawing and refreezing are major dangers for frozen food. Look for large ice crystals, solid areas of ice, discolored or dried-out food, or misshapen cartons or products. Reject any food that may have been thawed and refrozen.

Fresh Produce

Before tasting fruits and vegetables, wash them to remove contamination and insecticides. Also check for insect infestation. Train your employees not to damage produce by squeezing or pinching it.

Canned Foods

Never accept any home-canned foods. The risk for botulism is too great. Reject all cans with swollen sides or ends, flawed seals or seams, rust, dents, leaks, or foamy or bad smelling contents. Also reject any can without a label. *Never* taste a can's contents to test them—botulism can result from even one small taste.

Dry Goods

Dried fruits and vegetables, cereals and other grain products, sugar, flour, and rice must be received in dry, unbroken packaging. Dampness or mold may be signs of spoilage or bacterial growth. Holes or tears may be signs of pest infestation. Check flour or cereal for insects by sprinkling some of the contents on brown paper and watching for movement.

Rejecting Shipments

Your employees should know what to do whenever a shipment does not meet the standards of your food safety system.

If it is necessary to reject a shipment:

◆ Keep the unacceptable food from your other food and supplies.

◆ Tell the delivery person the exact problem with the food. Use your purchase agreement and documented standards to back up your case.

◆ Do not throw it out or let the delivery person remove it until a signed adjustment or credit slip is in hand.

◆ Record the incident in the log, including the food involved, the carton number if appropriate, the standard not met, and the type of adjustment made.

A Case in Point

On August 14, a number of foods were delivered to White Oaks Nursing Home during the lunch hour. The food products were all different—cases of frozen steaks, canned vegetables, frozen shrimp, fresh tomatoes, and fresh chicken for that night's dinner, and a case of potatoes.

Betty, the new assistant manager, thought that the best thing was to tell Ed, who was in charge of receiving, to put everything away and check it later. Ed asked her if it would be better to ask the delivery person to come back later. Since the chicken was for that evening's dinner, she did not think it was a good idea to delay delivery. Ed put the frozen steaks and shrimp in the freezer and the fresh chicken in the refrigerator. Then he put the fresh tomatoes, potatoes, and canned vegetables in dry storage.

"Whew," said Betty. "I'm glad that's over!" Then she went back to the front of the house.

What was done incorrectly?

What could be the possible result?

Answer to a Case in Point

Usually, later never comes. Therefore, we have to believe that the food at White Oaks Nursing Home went unchecked until each product was ready to be used. Even then, the cook or the server could be in such a hurry that the inspection may be short or not at all. As a result, a control point has been missed.

To prevent this from happening, Betty or her manager, needs to set up a time with each supplier to bring the food deliveries when business is slower. Ed was correct asking her to delay the delivery until they were not as busy, so he would have had time to inspect each shipment. Then, all rules for accepting or rejecting the foods could have been followed.

Remember, good purchasing and receiving procedures are a check of the safety and quality of food products. Luckily, there were no MAP or sous vide foods delivered in this shipment—otherwise a CCP would have been missed. The result could be foodborne illness.

In the next chapter, Angie, another employee at White Oaks Nursing Home, will prepare the chicken, unaware that it had been improperly received.

Chapter 6 Exercise

1. A reliable supplier is one who will:
   a. follow federal and local standards.
   b. always provide the lowest price.
   c. vary the time deliveries are made every week.
   d. provide give-aways with each new order.

2. When a shipment of food arrives, employees should:
   a. put everything away and inspect it later.
   b. inspect only the potentially hazardous foods.
   c. inspect all foods right away before storing them.
   d. stack it neatly on the dock and inspect it within 12 hours.

3. Game meats:
   a. should be delivered frozen in cartons.
   b. require suppliers who have followed governmental regulations.
   c. include cattle and goats.
   d. are always raised in the wild.

4. Reject any poultry that has:
   a. a product temperature lower than 40°F (4.4°C).
   b. green or purple blotches.
   c. a USDA stamp.
   d. been packed in self-draining crushed ice.

5. The internal temperature of shell eggs at receiving should be:

   a. 25°F (−3.9°C).
   b. 40°F (4.4°C).
   c. 55°F (12.8°C).
   d. 140°F (60°C).

6. In a package of frozen fish, a large solid mass of ice on the fish means it was:

   a. properly packed.
   b. more expensive.
   c. raised commercially.
   d. thawed and refrozen.

7. To measure the temperature of milk that is packaged in a soft bulk container:

   a. pour the milk into a bowl.
   b. use a TTI.
   c. poke a small hole in the container.
   d. fold the container around the thermometer.

8. If a supplier offers you a good deal on home-canned tomatoes:

   a. take it, but wash off the cans.
   b. reject it, but see if the supplier has home-canned green beans.
   c. reject it, or any other offer for home-canned foods.
   d. take it, but chill the tomatoes to 60°F (15.6°C).

9. Sous vide foods are foods that:

   a. are always safer than natural ingredients.
   b. must be purchased from licensed producers.
   c. contain less oxygen than other kinds of food.
   d. are cooked by broiling.

10. If it is necessary to reject a shipment:

   a. call the health department.
   b. temporarily store the unacceptable shipment.
   c. obtain a signed credit slip from your delivery person.
   d. do not record any information about this shipment.

CHAPTER 7: STORING FOOD SAFELY

Test Your Food Safety IQ

1. **True or False:** For efficiency, fill storage shelves as much as possible. (See *Cold Storage*, page 90.)

2. **True or False:** The temperature of each storage area should be checked regularly. (See *Procedures*, page 89.)

3. **True or False:** Use your storage freezer to cool food before refrigerating it. (See *Cold Storage*, page 90.)

4. **True or False:** The coldest area in a refrigerator is near the door. (See *Cold Storage*, page 90.)

5. **True or False:** Always wash eggs before refrigerating them. (See *Eggs*, page 94.)

Learning Objectives

After completing this chapter, you should be able to:

◆ Set up storage standards and procedures.

◆ Provide storage equipment and facilities.

◆ Use the different types of storage facilities appropriately.

◆ Store foods safely.

General Storage Guidelines

Provide the necessary facilities and equipment for storing food safely. Train your employees to monitor each kind of food while it is in storage.

Procedures

Employees who work in storage areas should be trained to:

◆ Use the *First In, First Out* (FIFO) method. On each package, either write the expiration date, when the item was received, or when it was stored after preparation. Shelve new supplies behind old, so the old are used first. Regularly check expiration dates.

◆ Measure and record storage area and stored food temperatures as part of your HACCP system. Remember that a thermostat measures the equipment's temperature and a thermocouple or bi-metallic stemmed thermometer measures the food's internal temperature.

♦ Follow the corrective action each time a food item has been time or temperature abused, passed its expiration date, or when a storage area is at the wrong temperature.

♦ Keep unauthorized persons out of storage areas.

♦ Clean up all spills and leaks and remove dirty packaging and other trash right away.

♦ Cleaning supplies and other chemicals should be stored:

  • In locked rooms or cabinets away from food preparation and storage areas.

  • In original containers or in sturdy containers labeled with the contents and their hazards.

  • Near *material safety data sheets* (MSDSs). These are written descriptions of the contents, hazards, and handling procedures for chemicals and products containing chemicals. They are required by OSHA and need to be readily available. (See *Chapter 10*.)

  • *Never* use empty chemical containers to store food and *never* put chemicals in used food containers.

Remember: Do not take chances with food. When in doubt, throw it out!

Cold Storage

Refrigerators, deep chilling units, and freezers are your main tools for keeping potentially hazardous foods cold enough to prevent bacteria from growing. To keep your cold storage units effective:

♦ Use storage units for storage only (see *Chapter 8* for cooling techniques).

♦ Monitor food temperatures.

♦ Avoid overloading, which taxes the cooling unit and cuts down air circulation.

♦ Keep the unit door shut as much as possible. Only open it for short time periods.

♦ Place thermometers in the warmest area (usually by the door) and the coldest area (usually in the back) of each unit. Some units also include a read-out panel outside the unit so you can check the inside temperature without opening the door.

Refrigerated Storage

Use refrigerators only to keep foods at an internal product temperature of 40°F (4.4°C) for short time periods. Temperatures lower than 32°F (0°C) may freeze foods and damage them.

◆ Store cooked and ready-to-eat foods above raw foods to avoid cross-contamination (see *Exhibit 7.1*). When storing raw foods, use the following top-to-bottom order based on end cooked product internal temperatures (see *Chapter 8* for final cooking tempertures):

- Cooked and ready-to-eat foods (*top shelf*)

- Raw fish

- Raw unground beef

- Raw pork, ham, bacon, and sausage

- Raw ground beef and ground pork

- Raw chicken (*bottom shelf*)

◆ *Never* line the shelves. It cuts down air circulation necessary for proper cooling.

◆ Ideally, use two refrigerators: one for meat, poultry, fish, and dairy products and another for fruits and vegetables. Another possibility is to use one unit for raw foods and one for cooked foods.

◆ If doors are not practical, use cooler curtains or plastic insulating strips in walk-in refrigerators.

EXHIBIT 7.1 PROPER REFRIGERATOR STORAGE

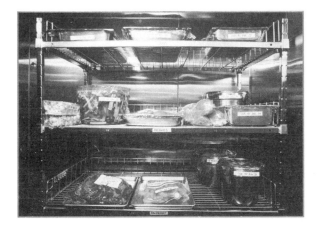

Deep Chilling Storage

Deep chilling involves storing food at unit temperatures of 26° to 32°F (–3.3° to 0°C) for short time periods. These temperatures limit bacterial growth without damaging food quality. Deep chilling is often used for sous vide products, poultry, meat, and seafood.

Use only specially designed units or refrigerators that have the capacity to reach and hold these temperatures. Many refrigerators are not able to do so and simply turning down the temperature setting will not help in the long run and could freeze the compressor coils.

Freezer Storage

Use freezer units only to store already chilled or frozen foods at a unit temperature of 0°F (–17.8° C) or lower.

Thawing and refreezing damages food quality. More importantly, food that has been thawed and refrozen is more likely to have been exposed to conditions that support bacterial growth.

◆ Regularly check unit and food temperatures. Foods that may be damaged by long freezing include hamburger, mackerel, salmon, bluegill, turkey, pork, creamed foods, sauces, custards, gravies, and puddings.

◆ Move frozen foods from receiving to freezer storage as soon as they are inspected.

◆ Regularly defrost units. Move frozen foods to another freezer during defrosting.

◆ *Never* refreeze thawed food until after it has been thoroughly cooked.

Dry Storage

Keep packages of dried fruits and vegetables, cereals and other grain products, sugar, flour, and rice intact and dry. These foods may be stored for long time periods, but water and high humidity may cause bacterial growth. The expiration date of each item should be checked.

◆ Store dry foods at least six inches off the floor and out of sunlight (see *Exhibit 7.2*).

◆ Storage area temperatures should be 50° to 70°F (10° to 21.1°C) with a relative humidity of 50 to 60 percent. If high humidity is a problem, try a dehumidifier.

◆ Dry storage areas must be well-ventilated and pest-free.

EXHIBIT 7.2 PROPER DRY STORAGE

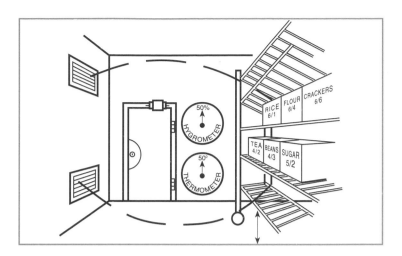

Storing Specific Foods

Your food safety system should include standards for storing each kind of food.

Meat

In general, store meat in the coldest section of the unit. Discard meat that develops a sour odor, discoloration, slime, or mold.

◆ Refrigerate most meats at an internal product temperature of 40°F (4.4°C) or lower. All refrigerated raw meats should be wrapped air-tight.

◆ Freeze most meats at a unit temperature of 0°F (–17.8°C). All frozen meats should be wrapped air-tight.

◆ Beef quarters and sides can be hung from sanitary hooks if nothing is stored above or below them.

◆ Ham, bacon, and luncheon meats should not be frozen unless they were delivered frozen.

Poultry

Store poultry in the coldest section of the unit. Discard any poultry that develops green or purple blotches, stickiness, or a sour odor.

◆ Refrigerate poultry at an internal product temperature of 40°F (4.4°C) or lower. Whole poultry should be used within three days of arrival. Parts and cooked poultry should be used within one to two days.

◆ Freeze poultry at a unit temperature of 0°F (−17.8°C). All frozen poultry should be wrapped air-tight.

◆ Raw chicken received packed in ice can be stored in that ice. Beds or chests of ice must be self-draining. Regularly change the ice and sanitize the container.

Eggs

Eggs are available in several forms. Each type has specific storage requirements:

◆ Shell eggs:

- Store in their original container at an internal product temperature of 40°F (4.4°C) or lower.

- Use FIFO.

- Take out only as many eggs as you need for immediate use.

- Do not wash shell eggs before storing them. They are washed and sanitized before they are packed.

- Do not *pool* (crack several into a bowl) eggs unless they are used immediately.

◆ Other Types of Eggs:

- *Frozen:* Store at a unit temperature of 0°F (−17.8°C). Thaw at refrigerator temperature.

- *Liquid and Refrigerated Egg Products:* Store in their original container and refrigerate at 40°F (4.4°C) or lower immediately after delivery. Always store in a refrigerator, keeping the product's seal intact.

- *Dried Egg Products:* Store in their original container in a cool, dry place away from light, preferably in a refrigerator. After opening, seal tightly for storage and refrigeration.

Fish and Shellfish

Store fish at an internal product temperature of 40°F (4.4°C) or lower.

◆ Fresh whole fish may be stored up to three days if delivered on crushed or flaked ice. Beds or chests of ice must be self-draining. Regularly change the ice and sanitize the container.

If not received on ice, the fish should be used within 48 hours. Fish fillets should be kept in air-tight, moisture-proof wrappings.

◆ Frozen fish should be stored at a unit temperature of 0°F (–17.8°C) or lower in air-tight, moisture-proof wrappings.

◆ Fish (other than shellfish) meant to be eaten raw must be received frozen or be frozen before being served. Freeze fish at:

- –4°F (–20°C) or lower for 168 hours (7 days) in a freezer, or

- –31°F (–35°C) or lower for 15 hours in a blast freezer.

◆ Store MAP fish at the specified temperature. Observe the expiration date listed on the package.

◆ Store live shellfish in their original containers at a unit temperature of 45°F (7.2°C).

The FDA's *1993 Food Code* states that molluscan shellfish (oysters, clams, mussels, and scallops) that are meant to be eaten cannot be kept in a display tank visible to the public. All tanks must carry a sign that states the shellfish are for display only.

To display shellfish that are meant to be eaten, a variance must be obtained from the health department and a HACCP-based plan submitted, showing that the:

- Water used in the display tank has not been used with any other fish.

- Safety and quality of the shellfish are not threatened.

- Shellstock identification tag is visible on the tank.

Dairy Products

Store dairy products at an internal product temperature of 40°F (4.4°C) or lower. Keep packages tightly covered and away from foods with strong odors, such as onions or fish. Keep only until the expiration or use-by date.

Ice cream may be stored at a product temperature of 6° to 10°F (–14.4° to –12.2°C).

Fresh Produce

Most whole raw fruits and vegetables can be stored at a unit temperature of 40° to 45°F (4.4° to 7.2°C). Whole raw fruits or vegetables and cut raw vegetables (such as celery or carrot sticks and potatoes) that were received packed in ice can be stored in that ice. Beds or chests of ice must be self-draining. Regularly change the ice.

Canned Goods

The storage area temperature for canned goods should be 50° to 70°F (10° to 21.1°C).

Modified Atmosphere (MAP) and Sous Vide Packaged Foods

Store these products at manufacturer recommended temperatures. Discard packages that are torn or past their expiration dates.

Refrigerated MAP foods can be stored for a maximum of 20 days if kept at their recommended temperature. Some jurisdictions require that MAP foods be used within 14 days of receiving. Check with your local jurisdiction for storage requirements.

Aseptically and Ultra-Pasteurized Packaged Foods

Ultra-pasteurized foods, such as milk products in cartons, have been heat-treated to kill disease-causing micro-organisms. Foods labeled UHT have been ultra-pasteurized (high temperature/short time) and aseptically packaged (hermetically sealed).

UHT-packaged foods may or may not be stored under refrigeration until they are opened. After being opened, they must be refrigerated. Ultra-pasteurized products that have not been aseptically packaged must be kept at an internal product temperature of 40°F (4.4°C) or lower throughout delivery, storage, preparation, service, and display.

Facilities and Equipment

Food must be kept out of the temperature danger zone (40° to 140°F [4.4° to 60°C]) and safe from all sources of contamination.

◆ Control the temperature in each storage area. Each area should be equipped with a hanging or built-in thermometer for measuring the ambient (surrounding) temperature. These thermometers should be accurate to ±2.7°F (±1.5°C).

◆ Let air circulate around food. Provide enough slatted shelves in storerooms, refrigerators, and freezers so packages do not have to be stacked on the floor, on top of each other, or against walls.

◆ Store food in original packaging as long as the packaging is clean, dry, and intact. *Never* reuse old wrappings or containers.

◆ Repackage food in leak-proof, pest-proof, non-absorbent, sanitary containers with tight-fitting lids.

- Store food only in proper storage areas. *Never* use locker rooms, restrooms, furnace rooms, hallways, stairwells, or garbage areas for storage.

- Keep food away from sewer and water lines, drains, and condensation dripping from pipes or ceilings.

- Clean and sanitize all utensils and equipment, such as carts, dollies, and delivery vehicles.

- Store fishing bait in a covered bait station away from stored food.

A Case in Point

At 2:00 PM on August 14, the foodservice employees at White Oaks Nursing Home were busy cleaning up from lunch and beginning to prepare for dinner. Pete, a kitchen assistant, put the still-hot container of left-over vegetable soup in the refrigerator to cool. Angie, a cook, began deboning the raw chicken breasts (see *A Case in Point* in *Chapter 6* to review important information about this chicken shipment) for the evening meal. Then she placed the uncovered chicken breasts in the same refrigerator where Pete had put the hot soup, being careful to place them on the top shelf away from the soup. Next, Angie iced a carrot cake she had baked in the morning. She put the iced carrot cake in the refrigerator directly below the chicken breasts.

What storage errors were made?

What foods are at risk?

Answer to a Case in Point

Since a refrigerator is designed to receive cold foods and maintain their temperature, Pete should have cooled the vegetable soup before putting it in the refrigerator. His action may have raised the refrigerator's temperature, creating an unsafe environment for the raw chicken breasts. We know from *A Case in Point* in *Chapter 6* that the chicken had already been exposed to improper temperatures.

The temperature of a refrigeration unit can rise from 39° to 55°F (3.9° to 12.8°C) whenever hot leftovers are placed in the unit. This danger can be avoided by dividing large quantities of food into shallow pans and using rapid quick-chilling techniques, such as ice baths or specially designed equipment like quick-chilling refrigerators to cool food. Quick-chilling refrigerator systems are designed to hold shallow pans and smaller units of foods for fast chilling before putting the food in regular refrigeration units. These units should be used as interim cooling systems and not as cold storage refrigerators.

The type and capacity of a refrigerator are important. Most refrigerators are cold holding units designed to receive food at 40°F (4.4°C) or lower and keep them cold. They do not have the capacity to receive large amounts of hot food and cool them. Special refrigerators may be necessary for operations that cool large amounts of hot food on a regular basis.

Lastly, Angie placed the ready-to-eat carrot cake directly below the raw chicken breasts. There is a strong chance that fluid from the raw chicken could splash onto the carrot cake when the chicken breasts are removed from the refrigerator. The chicken should have been stored on the bottom shelf, below the cake.

Chapter 7 Exercise

1. Locker rooms, restrooms, furnace rooms, and stairwells are:

   a. unacceptable as food storage areas.
   b. acceptable food storage areas for less than three days.
   c. acceptable food storage areas if kept clean.
   d. unacceptable food storage areas, but a hallway may be used.

2. Under the FIFO method, foods are used:

   a. in relation to size.
   b. in the order in which they were received.
   c. by selecting the newest foods first.
   d. by the cost of the food.

3. If stored foods have passed their expiration date:

   a. freeze the food for later use.
   b. cook and serve the food at once.
   c. discard the food.
   d. leave it on the shelf for later use.

4. If you have hot food that you want to refrigerate:

   a. first cool it in the freezer.
   b. put it in the refrigerator to cool.
   c. leave it out on the stove overnight.
   d. cool it, then refrigerate it.

5. Shelves in a refrigerator should be:

   a. lined with aluminum foil.
   b. slatted.
   c. heavily loaded for maximum use.
   d. covered in plastic wrap.

6. To deep chill and store foods at a unit temperature of 26° to 32°F (−3.3° to 0°C):

   a. use only units that are designed to reach and hold these temperatures.
   b. turn the setting on your refrigerator to its lowest point.
   c. use your freezer to first cool them.
   d. put them unwrapped in the refrigerator.

7. While you are defrosting your freezer, put the frozen food in:

    a. the refrigerator.
    b. another freezer.
    c. a microwave to thaw.
    d. the sink.

8. The coldest section of the refrigerator is *most* likely to be near the:

    a. door.
    b. door racks.
    c. top.
    d. back.

9. Which one of the following should you do when handling shell eggs?

    a. Store them at a unit temperature of 60°F (15.6°C).
    b. Wash them before putting them into the refrigerator.
    c. Crack a number of them into a large bowl for later preparation.
    d. Take out only as many as you need for immediate use.

10. Frozen raw fish should be stored:

    a. unwrapped.
    b. at 30°F (–1.1°C) for long-term thawing.
    c. wrapped in air-tight and moisture-proof wrapping.
    d. near the freezer door.

CHAPTER 8: KEEPING FOOD SAFE DURING PREPARATION AND SERVICE

Test Your Food Safety IQ

1. **True or False:** Thawing food at room temperature is safe. (See *Methods for Safe Thawing*, page 101.)

2. **True or False:** Cook foods no higher than the recommended internal temperature. (See *Cooking*, page 102.)

3. **True or False:** Microwaved meats should be allowed to stand after cooking. (See *Microwaved Meats*, page 102.)

4. **True or False:** Prepare meat and poultry salads 48 hours before serving them. (See *Protein Egg Salads and Sandwiches*, page 104.)

5. **True or False:** Batters and breadings containing eggs are reusable if refrigerated. (See *Batters and Breadings*, page 105.)

Learning Objectives

After completing this chapter, you should be able to keep food safe throughout:

- ◆ Thawing.
- ◆ Preparing.
- ◆ Cooking.
- ◆ Holding.
- ◆ Serving.
- ◆ Cooling.
- ◆ Reheating.

Time-Temperature Principle

Many foods are most at risk during preparation and service. As foods are thawed, cooked, held, served, cooled, and reheated, they may pass several times through the temperature danger zone of **40° to 140°F (4.4° to 60°C)**. Each time food is handled, it runs the risk of cross-contamination from other food and from food-contact surfaces, such as human hands, cutting boards, and utensils.

Prevent potentially hazardous foods from spending more than four hours total in the temperature danger zone.

REMEMBER: KEEP HOT FOOD HOT AND COLD FOOD COLD!

Methods for Safe Thawing

Thaw food only by these four methods:

1. In a refrigerator.

   ◆ Store raw foods on the lowest shelves to prevent them from dripping or splashing on other foods.

   ◆ Allow a day or more for large items, such as turkeys and roasts, to thaw.

   ◆ Carefully use the slacking process. *Slacking* involves allowing food to gradually warm from frozen to unfrozen so that it cooks more evenly. For example, you might allow a large block of frozen spinach to warm from –10°F (–23.3°C) to 25°F (–3.9°C). Slacking frozen foods should be done just before cooking, and the food must become no warmer than 40°F (4.4°C).

2. Under running potable (drinkable) water at a temperature of 70°F (21.1°C) or lower. The product should be thawed within two hours, then prepped, and cooked.

   ◆ Use a large cleaned and sanitized sink used only for thawing.

   ◆ Use a stream of water strong enough to wash off loose particles of skin or dirt. Do not let the water splash on other food or food-contact surfaces.

   ◆ Remove the food from the sink as soon as it is thawed. Sanitize the sink and all utensils used in thawing.

   Note: This method does not work for turkeys and large cuts of meat.

3. As part of the cooking process. This method works well with vegetables, seafood (such as shrimp), hamburger patties, pie shells, and similar foods—but not with large items. Allow longer than normal cooking time because the items are frozen.

4. In a microwave. Use this method only if the food will be moved immediately to other cooking equipment or finished immediately in the microwave. This method is not effective for large items.

Preparing Food for Cooking

Detailed recipes, time and temperature controls, and sanitary procedures are the keys to safety.

◆ Use properly cleaned and sanitized utensils and practice good personal hygiene.

◆ Use recipes that specify fat content, size, and thickness of each portion—this helps predict cooking time.

◆ Refrigerate foods before preparation.

◆ Prepare small batches of food no farther in advance than necessary. Return them to the refrigerator before cooking or serving.

◆ Use sanitized cutting boards and knives to avoid cross-contamination.

◆ Wash fruits and vegetables in sinks used only for food preparation.

Cooking

Recipe instructions for cooking should specify cooking times and end product internal temperatures. Employees should:

◆ Cook foods to higher than their minimum safe internal temperatures if their quality will not be compromised (see *Exhibit 8.1*).

◆ Measure food temperatures with a thermocouple or thermometer, accurate to ±2°F or ±1°C. *Never* rely on a "best guess," "experience," or equipment (such as oven) thermometers.

◆ Measure internal food temperatures in several places. Clean and sanitize thermometers before and after each use. (See *Using Thermometers* in *Chapter 6*.)

◆ Avoid overloading cooking surfaces and ovens. The unit's temperature may drop or foods may spill on each other.

◆ Allow the temperature of cooking equipment to return to required temperatures between batches.

◆ Correctly taste foods. (See *Tasting Food During Preparation* in *Chapter 3*.)

Cooked Food Temperature Requirements

For each of the foods listed in *Exhibit 8.1*, internal end product temperatures and safe procedures are suggested (see *Exhibit 8.1*).

Some high-protein foods require special handling (see *Exhibit 8.2*).

Microwaved Meats

All raw animal foods that are cooked with a microwave should be covered to stay moist. Use the following guidelines for cooking:

◆ Add a minimum 25°F or 14°C to each of the above internal cooking temperatures. For example, the internal cooking temperature of chicken cooked in a microwave would be 190°F (87.8°C) or higher.

◆ Rotate or stir midway through cooking to help spread the heat.

◆ Let stand for 2 minutes after cooking so that all parts of the meat heat to the required internal temperature.

EXHIBIT 8.1 MINIMUM SAFE INTERNAL COOKING TEMPERATURES

Poultry, stuffing, stuffed meat, and stuffed pastas—165°F (73.9°C) for 15 seconds; cook stuffing and meat first, then stuff the food

Ground meats (including ground beef and ground pork)—155°F (68.3°C) for 15 seconds

Pork, game animals, comminuted fish and meats, injected meats, and eggs in multi-serving batches—155°F (68.3°C) for 15 seconds; 150°F (65.6°C) for 1 minute; or 145°F (62.8°C) for 3 minutes

Beef roasts—145°F (62.8°C) for 3 minutes; 140°F (60°C) for 12 minutes; or 130°F (54.4°C) for 121 minutes

Fish, seafood, beef (cubes, slices, etc.), veal, lamb, mutton, shell eggs for immediate service for a customer's order, and all other potentially hazardous foods not listed in this exhibit—145°F (62.8°C) for 15 seconds

EXHIBIT 8.2 MINIMUM SAFE INTERNAL COOKING TEMPERATURES FOR SPECIAL MEATS AND MEAT PRODUCTS

GAME MEATS:

Field-dressed game—165°F (73.9°C) for at least 15 seconds

Commercially dressed game—155°F (68.3°C) for at least 15 seconds

OTHER RAW ANIMAL FOODS:

Comminuted foods (chopped, ground, flaked, or minced foods, such as ground beef, gyros, sausage, and fish)—155°F (68.3°C) for at least 15 seconds

Injected Meats—155°F (68.3°C) for at least 15 seconds

Eggs and Egg-Based Mixtures

To prepare eggs safely:

◆ Do not pool eggs if they will not be prepared in small batches and used immediately.

◆ Do not stack egg trays near the grill.

◆ Do not allow shells to touch or mix with egg contents.

◆ Do not use processing equipment that grinds entire eggs and separates out the shells.

◆ Use pasteurized eggs in all recipes:

• In which eggs are not cooked or cannot be cooked to 145°F (62.8°C) or higher. This includes meringues and mousses, Caesar salad dressings, hollandaise and bernaise sauces, eggnog, and mayonnaise.

• Served to the elderly, ill, pregnant women, infants, and other diners with weakened immune systems.

◆ Use cleaned and sanitized bowls, whisks, blenders, and other utensils for each new order or batch.

Cook eggs to 145°F (62.8°C) for at least 15 seconds. Cook whole eggs until the white is completely set and the yolk begins to thicken. Cook scrambled eggs and omelets until they are firm and no liquid egg is visible.

Protein Egg Salads and Sandwiches

Salads and sandwiches containing meat, poultry, eggs, or fish require careful handling. This is especially true when you make large quantities or prepare these items several hours before service.

◆ Use properly cleaned and sanitized utensils and practice good personal hygiene.

◆ Prepare pasta, meat, egg, and fish salads less than 24 hours before service.

◆ Chill all sandwich and salad ingredients (including bread) to 40°F (4.4°C) or lower before making the meal.

◆ Wash all vegetables and fruits. Blanch items such as celery and carrots.

◆ Use commercially made mayonnaise, not homemade.

◆ Make only small batches for large orders. Refrigerate each batch before service.

Stuffings

Stuffings insulate the food they fill which prevents thorough cooking. When cooking stuffed foods made entirely or partly of potentially hazardous foods, you should:

♦ Separately cook stuffing to a minimum internal temperature of 165°F (73.9°C) for at least 15 seconds. If the cooked stuffing is not used immediately, it should be cooled and then refrigerated (see *Cooling Procedures* later in this chapter).

♦ Cook stuffed meats, fish, pasta, and poultry to 165°F (73.9°C) or higher for at least 15 seconds.

Batters and Breadings

Many of the foods cooked in batter or breading, such as poultry, fish, and shellfish, are potentially hazardous. Batters made with eggs are also potentially hazardous. Used as stuffings, batters and breadings may insulate the food they cover and prevent complete cooking.

To avoid this hazard, consider using commercially made battered and breaded items. Many pre-prepared frozen items can also be cooked frozen.

If you make and use your own batters and breadings, you should:

♦ Use pasteurized eggs, rather than shell eggs.

♦ Refrigerate the ingredients for batters and breadings, as well as the items to be coated.

♦ Make small amounts of breadings and batters. Coat only small batches of food at a time. After coating, return the food to the refrigerator.

♦ Thoroughly cook items to be sure that the coatings do not keep out the heat. For deep frying, you should:

• Match the batch size to your deep fryer's capacity—do not overload.

• Keep the oil temperature constant. Allow time between batches for the oil to return to the required temperature.

• Record the cooking time for each type of food and add it to your recipe and flowcharts.

♦ Do not reuse batter in which items have been dipped.

Holding

General Procedures

♦ Make only small batches. Breaded, fried, and baked foods should be held only for very short time periods.

♦ Regularly stir held foods and measure their temperatures with a thermocouple or bi-metallic stemmed thermometer every two hours. Do not rely on the thermostat on the holding equipment—that measures the temperature of the equipment, not of the food. Record temperatures in a log.

♦ Use covered holding pans and provide long-handled spoons or tongs so human hands do not touch the food. Place spoons and tongs in the food with their handles pointed toward the user or store these utensils in drinkable running water.

♦ *Never* mix new food with old or raw food with cooked.

♦ Use properly cleaned and sanitized utensils and practice good personal hygiene.

Holding Hot Food

To provide control during holding, you should:

♦ Use only hot-holding equipment that can keep foods at 140°F (60°C) or higher. Holding equipment includes steam tables, double boilers, bain maries, heated cabinets, and chafing dishes.

♦ *Never* use hot-holding equipment to cook or reheat food—only to keep food hot.

♦ Measure food temperatures every two hours. Record them in a log.

Holding Cold Food

Cold cooked and raw foods must also be kept safe from temperature abuse and contamination. To provide control during holding, you should:

♦ Use only cold-holding equipment that can keep foods at 40°F (4.4°C) or lower.

♦ Hold ready-to-eat cold foods in pans or on plates, *never* directly on ice. Be sure ice used to surround chilled foods drains away from the food. Drip pans should be sanitized after each use.

♦ Measure food temperatures every two hours. Record them in a log.

Serving

Set up serving procedures to help employees safeguard food and customers (see *Exhibit 8.3* and *Chapter 3*). Stagger employee assignments so that an employee does not serve food, set tables, and clear dirty dishes during a single shift.

Train employees to:

◆ Use properly cleaned and sanitized utensils and to practice good personal hygiene.

◆ *Never* touch the food-contact areas of glasses, cups, plates, and tableware.

◆ *Never* stack cups or bowls before serving them—the bottom of one will touch the rim of the one below it and possibly contaminate it.

◆ Use plastic or metal tongs or scoops to get ice—*never* use glass that may break in the ice.

Safeguarding Customers

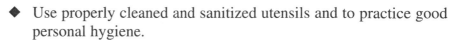

Prepare in advance for customers' requests and problems and train your employees to respond to customers' needs with care. For example, if a customer with a food allergy asks if a certain ingredient is in one of the menu items, an employee should be able to answer the question or ask the chef or cook for the answer. The employee needs to reply to the customer honestly, even if it is an, "I'm sorry, but I asked and we really aren't sure. Although the chef did suggest you try . . ." At no time should the employee ignore a customer's question or give an answer that may be wrong.

EXHIBIT 8.3 PROPER SERVING

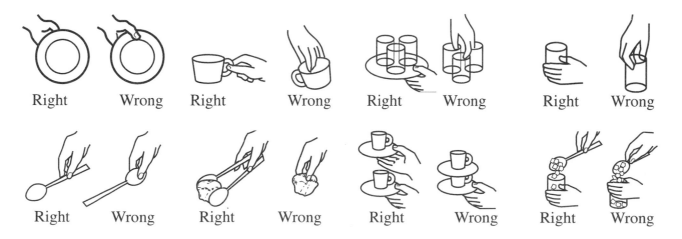

| Right | Wrong | Right | Wrong | Right | Wrong | Right | Wrong |

| Right | Wrong | Right | Wrong | Right | Wrong | Right | Wrong |

Cooling

Food must be cooled to 40°F (4.4°C) or lower in less than four hours total after cooking or hot holding. Many state and local health departments use other codes that permit 45°F (7.2°C) or lower in less than four hours.

The FDA's *1993 Food Code* recommends cooling in two stages:

1. From 140°F (60°C) to 70°F (21.1°C) in two hours, and

2. To 41°F (5.0°C) in four hours for a total of six hours cooling time.

Cooling Procedures

When cooling large pieces or batches of hot food, begin by cutting large items into smaller pieces or dividing large batches into several smaller ones (see *Exhibit 8.4*). Place the smaller amounts in pre-chilled stainless steel pans.

Then, use one of the following methods:

◆ Place the pans in larger pans of ice. Stir foods as they cool. Then:

    • Put the food in shallow stainless steel pans on the upper shelves of a refrigerator. Thick foods, such as chili and stew, should be in pans with a product depth no more than two inches. Thinner liquids, such as broth and soup, may be in pans with a product depth no more than three inches deep.

    • Place the pans so air circulates around them. Put uncovered pans on top shelves to cool further, then cover them loosely to maintain air flow. (Some health departments require pans to be covered at all times.)

    • If food is not cooled to 40°F (4.4°C) after four hours, take corrective action. Reheat it to 165°F (73.9°C) for at least 15 seconds in two hours. If food is not served immediately, discard it.

    • Label properly cooled and stored foods with the date and time they were prepared. If food is not used within 10 days, discard it.

◆ Place the pans in a quick-chill unit, tumbler chiller, or cold-jacketed kettle. *Never* use storage refrigerators or freezers to cool foods. Hot foods can raise the temperature of the unit and endanger the other foods stored there.

Record the cooling times required for each type of food and add these times to your recipes and flowcharts.

EXHIBIT 8.4 COOLING FOODS

Reduce Size

Shallow Pans

Ice-Water Bath

Blast Chiller

Reheating

To keep reheated food safe:

◆ Reheat all previously cooked food to an internal temperature of at least 165°F (73.9°C) for at least 15 seconds within two hours. If the food cannot be reheated within two hours, discard it.

◆ Food reheated in a microwave must be heated to at least 190°F (87.8°C). Let the food stand for two minutes after reheating so the heat spreads evenly throughout all parts of the food.

◆ Ready-to-eat food taken from a commercially processed, hermetically sealed container or from an intact package from a processing plant must be reheated (often called rethermalized) to 140°F (60°C) or higher.

◆ Transfer reheated food to holding equipment only when the food is at 165°F (73.9°C).

◆ Use cooking ranges, ovens, steamers, and microwaves to reheat food—never use hot-holding equipment because it is not designed to reach and maintain the necessary temperatures.

◆ Reheat food in small batches to shorten reheating time.

◆ Reheat food only once.

Record the reheating times for each type of food and add these times to your recipes and flowcharts.

Ten Rules of Safe Foodhandling

This list is in the order of the flow of food.

1 Require strict personal hygiene from all employees.

2 Identify all potentially hazardous foods on your menu and write out your foodhandling procedures. Make these written procedures part of employee training, everyday tasks, and regular self-inspection.

3 Obtain foods and other supplies from reputable, approved sources.

4 Observe the rules for time and temperature and for preventing cross-contamination in storing and handling food prepared in advance of service.

5 Keep raw products separate from ready-to-eat foods.

6 Avoid cross-contamination of raw and ready-to-eat foods from hands, equipment, and utensils. Clean and sanitize food-contact surfaces and equipment before and after every use, after an interruption, and at least every four hours during continual use.

7 Cook or heat-process food to above the recommended minimum temperature.

8 Keep hot foods hot and cold foods cold. Store or hold foods at **140°F (60°C) or higher or at 40°F (4.4°C) or lower**

9 **Chill** cooked food to **40°F (4.4°C) within four hours** (see *Cooling Procedures* in this chapter).

10 **Reheat** food to an internal temperature of at least **165°F (73.9°C) for at least 15 seconds within two hours**.

A Case in Point

During the evening of August 14, Angie prepared the chicken breasts (see *A Case in Point* in *Chapters 6* and *7*) for dinner, a favorite among the residents of White Oaks Nursing Home. Because several residents were away from the home and Angie had not been informed of this, she prepared more chicken breasts than were necessary.

"No problem," Angie thought. "I'll just use the leftover chicken to make chicken salad. I can serve that for lunch later this week." She refrigerated the leftover chicken in a large container at an air temperature of 40°F (4.4°C) until she was ready to prepare lunch on the following Monday.

Next Monday morning, Juanita, the manager, observed Angie preparing chicken salad, fresh fruit, cole slaw, and tapioca pudding for lunch.

"Aren't those chicken breasts leftover from last week?" Juanita asked. When Angie nodded, Juanita asked her to stop preparing lunch.

Why did Juanita stop Angie?

Answer to a Case in Point

Juanita stopped Angie because the chicken was not safe to serve. The chicken was already at risk when it was served an entire week ago, because it had been both improperly received and stored. Fortunately, White Oaks Nursing Home did not experience any foodborne illness when the chicken was served at that time. If the nursing home's foodservice manager had provided proper flowcharts and recipes, Angie would have known how to safely prepare and store leftover food items.

Chapter 8 Exercise

1. Which one of the following is required for safe foodhandling?

    a. Store hot foods at 40°F (4.4°C) or higher.

    b. Mix raw food with ready-to-eat food.

    c. Clean and sanitize food-contact surfaces only at the end of the day.

    d. Cook food to at least the recommended internal temperature.

2. Which one of the following is the safest way to thaw a large food item, such as a 20-pound turkey?

    a. In a refrigerator.

    b. Under drinkable running water at a temperature of 70°F (21.1°C).

    c. At room temperature.

    d. Quickly in a microwave oven.

3. If you must prepare a large batch of ham sandwiches for later service, you should make:

    a. them all at one time, then refrigerate them.

    b. them all at one time, then cover and leave them on the counter.

    c. several at a time, then cover and refrigerate them.

    d. several at a time, then cover and leave them on the counter.

4. Which one of the following would you do if you were preparing a recipe using eggs that are not cooked to 145°F (62.8°C)?

    a. Use pasteurized eggs.

    b. Completely drop the menu item, even if it is your most popular.

    c. Use only Grade AA eggs.

    d. Chill the shell eggs to 40°F (4.4°C) before using.

5. Meats cooked in a microwave should be:

    a. cooked 25°F or 14°C lower than the minimum safe internal temperature.

    b. cooked 25°F or 14°C higher than the minimum safe internal temperature.

    c. left uncovered and turned once midway through cooking.

    d. not rotated or stirred during cooking.

6. Which one of the following is the correct way to cook a stuffed turkey?

    a. Cook the stuffing and the turkey together to 70°F (21.1°C).

    b. Separately cook the stuffing to 145°F (62.8°C).

    c. Separately cook the stuffing to 165°F (73.9°C).

    d. Cook the stuffing and the turkey together to 145°F (62.8°C).

7. A hot potentially hazardous food should be held at:

    a. 70°F (21.1°C).

    b. 120°F (48.9°C).

    c. 140°F (60°C).

    d. 165°F (73.9°C).

8. Which one of the following procedures should employees do when serving?

    a. Use bare hands to touch cooked food or food that will not be cooked.

    b. Touch the food-contact or mouth-contact parts of dishes and utensils.

    c. Stack cups or bowls before serving them.

    d. Use plastic or metal tongs or scoops to get ice.

9. Large amounts of *thick* food, such as chili, should be cooled in shallow pans with a product depth no greater than:

    a. 2 inches.

    b. 3 inches.

    c. 4 inches.

    d. 5 inches.

10. Which one of the following procedures should you do when reheating food?

    a. Reheat food to at least 165°F (73.9°C) for at least 15 seconds within two hours.

    b. Reheat food no more than three times.

    c. Reheat day-old beef soups and stews to 155°F (68.3°C) for at least 15 seconds`.

    d. Reheat casseroles in hot-holding equipment.

CHAPTER 9: SANITARY FACILITIES AND EQUIPMENT

Test Your Food Safety IQ

1. **True or False:** Always have building and remodeling plans reviewed by the health department. (See *Review of Building and Remodeling Plans*, page 113.)

2. **True or False:** Ovens should be placed against painted plaster walls. (See *Wall Materials*, page 116.)

3. **True or False:** Bad plumbing is no longer a danger to restaurants. (See *Plumbing*, page 123.)

4. **True or False:** Opening doors and windows to let fresh air into a kitchen is a good idea. (See *Ventilation*, page 126.)

5. **True or False:** Garbage containers need to be regularly cleaned and sanitized. (See *Garbage Disposal*, page 126.)

Learning Objectives

After completing this chapter, you should be able to:

◆ Describe a well-designed restaurant.

◆ Select proper equipment.

◆ Review utilities, lighting, and ventilation.

◆ Arrange for careful handling of garbage and solid waste.

Review of Building and Remodeling Plans

Local health departments usually review applications for new operating permits and for permit renewals. These agencies may also require submitting building plans before constructing a new building or remodeling.

Contact your health department for the specific site requirements. In most cases, plans should cover the following:

◆ Menu items, seating capacity, and volume of food to be served.

◆ Proposed *layout*, which is the order of equipment, work areas, and furniture, for dining and back-of-the-house areas. Describe how you will meet federal employee and customer access requirements in accordance with the Americans with Disabilities Act (ADA).

◆ Requirements for utilities, mechanical systems, and construction materials.

- Proposed equipment, including types, manufacturers, model numbers, locations, dimensions, performance capabilities, and installation requirements.

- Written standard procedures for operating in new or remodeled buildings.

Several agencies, such as the fire department or the Occupational Safety and Health Administration (OSHA), may be involved in the review. Add their recommendations to your plan.

Interior Construction, Design, and Materials

Designing for food safety means that every inside surface in your restaurant should be cleanable. *Clean* means free of visible soil and food waste, and *cleanable* means that soil and waste can be effectively removed by normal cleaning methods. The layout of your kitchen must not allow or cause food to be contaminated during preparation.

Workflow Patterns

Workflow refers to the order of the tasks to prepare a food item, beginning in the receiving area and leading to the dining room. The goal in designing a floor plan is to control safety, and quality at all critical points while keeping your production costs as low as possible. Keep the following principles in mind:

- Plan tasks so employees travel the shortest distance in the least time.

- Arrange tasks so employees do not criss-cross the work area or back-track. Avoid difficult patterns that can cause falls, collisions, and spills.

- Shorten trips to storage. Bring out food when needed, prepare it, and serve, hold, or return it to storage for later use.

- Have work spaces and equipment ready when food is brought out of storage. Complete the task, move the food to the next step, and clean and put away the equipment.

Dry Storage

Entrances should have tight-fitting, self-closing solid or screened doors. Floors should be as described below. (See *Chapter 7* for more information.)

- Walls should be covered with materials such as epoxy or enamel paints, stainless steel, or glazed tile.

- Shelves and table tops should be corrosion-resistant (still cleanable after repeated contact with food, water, and cleaning compounds) metals. Shelves should be slatted and wide enough to avoid overcrowding and loss of air circulation.

Shelf racks should be at least six inches away from the wall and off the floor for easier cleaning and to keep out pests and moisture.

◆ Bins for dry ingredients should be corrosion-resistant metal or food-grade plastic. Covers should keep out pests and moisture. Label bins with contents and use-by dates.

◆ Storage area windows should have frosted glass to keep sunlight from harming food quality.

◆ Steam pipes, ventilation ducts, water lines, and conduits should not be exposed.

Restrooms

Restrooms are usually regulated by local laws. (See *Chapter 3* for more information on handwashing facilities.) If possible, have separate restrooms for employees and customers. To keep food safe, customers must never pass through food preparation areas to get to or from restrooms. Supply restrooms with the following materials:

◆ Toilet paper, soap, and disposable towels, or air-blowing hand dryers. Common towels are not allowed. Pull-down towel holders may be provided only for customer use.

◆ A covered waste can for disposable paper towels and a separate one for feminine sanitary products.

Flooring

Flooring should be able to withstand the shock of constant traffic and items being dropped. It should also be hard for liquids to soak into the surface.

Kitchen, Storage, and Work Area Flooring

Flooring should repel liquids, be nonskid, and withstand strong cleaners.

◆ Use materials such as marble, terrazzo, quarry, or asphalt tiles. Seamless concrete can be used if regularly treated with chemical sealants (but not paint).

◆ Replace or repair all cracked or chipped flooring.

◆ Install coving with materials such as marble, terrazzo, quarry, or asphalt tiles, or sealed concrete. *Coving* is a curved, sealed edge between the wall and the floor. It removes sharp corners and gaps that make cleaning difficult. Remaining gaps should be smaller than $1/32$ of an inch (see *Exhibit 9.1*).

EXHIBIT 9.1 COVING

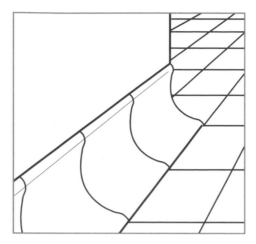

Dining Areas

Choose carpeting with a tight weave for easy cleaning. Vacuum daily and shampoo regularly.

Walls and Ceilings

In food preparation areas, walls and ceilings should repel liquids, be cleanable, smooth, and a light color to reflect light and reveal dirt.

Wall Materials

◆ *Ceramic tile:* Usable in almost all work areas. Grouting between tiles must be smooth, intact, and repel water and grease.

◆ *Stainless steel:* Usable in high-humidity food preparation areas where wear-and-tear is high.

◆ *Painted plaster or cinder block:* Usable in dry areas if sealed with nonflaking, soil-resistant, glossy paints, such as epoxy or acrylic enamel. Plaster and block are not recommended for areas where food and grease can splash walls, such as behind a stove.

Dining area drapes, wallpaper, and sconces (wall-mounted light fixtures) must be cleanable.

Ceiling Materials

Cleanable, sound-absorbing materials, such as smooth sealed plaster or plastic-covered tiles or panels, are recommended. Studs, joists, rafters, and pipes should not be exposed in the dining area unless finished and sealed for easy cleaning.

Food Preparation and Storage Equipment

Choose only equipment that meets industry and regulatory standards. Check equipment evaluations published by NSF *International*, formerly the National Sanitation Foundation, and by Underwriters Laboratories, Inc., (UL) (see *Exhibit 9.2*). Look for NSF and UL seals. *Never* use equipment intended for the home.

NSF standards require:

◆ Food-contact and food-splash surfaces that are:

  • Easy to reach.

  • Easily cleanable by normal methods.

  • Nontoxic, nonabsorbent, corrosion-resistant, nonreactive to food or cleaning products, and that do not leave a color, odor, or taste with food.

  • Smooth and free of pits, crevices, inside threads and shoulders, ledges, and rivet heads.

◆ Nontoxic lubricants.

◆ Rounded, tightly-sealed corners and edges.

◆ Solid and liquid waste traps that are easy to remove. For example, drink dispensers should have trays that can be removed without tools for cleaning or repair.

EXHIBIT 9.2 NSF *INTERNATIONAL* AND UL SYMBOLS

Courtesy of NSF International, *Ann Arbor, Michigan*   *Courtesy of Underwriter's Laboratories, Northbrook, Illinois*

Refrigerators and Freezers

Types of foodservice refrigerators and freezers are walk-ins and reach-ins, under-the-counter units, open units, and display units.

Required Safety Features For All Refrigerators and Freezers

Check for:

◆ Stainless steel construction or a combination of stainless steel and aluminum outside parts and plastic or galvanized liners (see *Exhibits 9.3* and *9.4*).

◆ Durable, easy-to-close doors with fixed or removable gaskets for easy cleaning.

◆ A system for draining water to the outside of the unit. (See *Plumbing* later in this chapter.)

◆ Enough cooling power for the stored food.

◆ Adequate lighting to see items and read labels.

◆ Interior edges and corners that are rounded and smooth.

◆ Enough slatted shelves or racks to avoid overcrowding and loss of air circulation.

◆ Cooling coils located so condensation will not drip on food or food-contact surfaces.

◆ Surfaces that are:

• Easy to reach, removable without tools, and easily cleanable by normal methods.

• Nontoxic, nonabsorbent, corrosion-resistant, nonreactive to food or cleaning products, and that do not leave a color, odor, or taste with food.

• Smooth and free of pits, crevices, inside threads and shoulders, ledges, and rivet heads.

Units can be purchased with alarms that sound when the unit loses power.

EXHIBIT 9.3 WALK-IN REFRIGERATOR

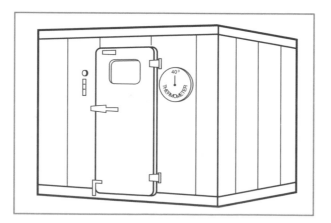

EXHIBIT 9.4 REACH-IN FREEZER

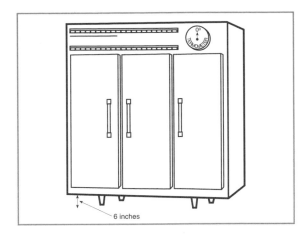

6 inches

Cook-Chill Equipment

Cook-chill equipment is designed to rapidly cool and then reheat foods. Large-scale foodservice operations most often use this equipment.

Blast Chillers

These units look like refrigerators and freezers with control panels (see *Exhibit 9.5*). Blast chillers chill foods from 140°F (60°C) to 37°F (2.8°C) in 90 minutes or less. Each unit should have an alarm or signal indicating the end of the chilling cycle.

Tumbler Chillers

These units have automated systems that combine steam-jacketed kettles, pump/filler stations, conveyors, and cooking and chilling tanks (see *Exhibit 9.6*). Tumbler chillers are often used with liquid or thick foods and can serve as reheating units.

EXHIBIT 9.5 BLAST CHILLER

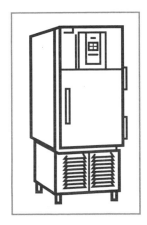

EXHIBIT 9.6 TUMBLER CHILLER

Courtesy of Groen, A Dover Industries Company, Elk Grove Village, Illinois

Cutting Boards

Separate cutting boards should be used for raw and cooked foods. Wash, rinse, and sanitize cutting boards after each use, after each food, after an interruption, and before four hours of continued use.

◆ Food-grade seamless hard rubber or acrylic blocks are safer and are required in some areas (see *Exhibit 9.7*). These boards can be cleaned and sanitized in a three-compartment sink or a dishwasher. Those that are too large to move also must be cleaned and sanitized.

◆ If wooden cutting boards are used, choose those made from seamless nontoxic hardwoods, such as maple, that do not leave an odor or taste with the food.

Be aware that bacteria can survive and grow in the cuts and scratches on wooden and plastic cutting boards. It is recommended that cutting boards be replaced or resurfaced on a regular basis or when the board has a number of cuts, scratches, or cracks.

Dishwashing Machines

The two most common dishwashing machines are high-temperature and chemical-sanitizing. Hot-water machines have higher energy costs, but chemical units require more drain board space for air-drying.

EXHIBIT 9.7 PROPER CUTTING BOARD

High-Temperature Dishwashers

◆ *Single-tank units* with stationary rack and outer doors. Dishes rest in a rack that does not move and are washed by detergent spray then rinsed.

◆ *Conveyor units* in which a moving belt carries dishes through wash, rinse, and sanitize cycles.

◆ *Carousel* or *circular units* with multiple tanks. Dishes are washed on moving belts or racks.

◆ *Flight-type units* in which dishes are carried on a conveyor belt. This large-capacity unit may have a built-in dryer and is most often used in institutions and very large foodservice operations.

Chemical Sanitizing Dishwashers

◆ Stationary rack units

• *Batch-Type/Dump:* Combines timed wash and rinse cycles in one tank. Detergent and sanitizing solutions are automatically dispensed. The tank is dumped (emptied) after each cycle.

• *Recirculating Door-Type/Non-Dump:* Does not dump water between cycles. Wash water is diluted with fresh water and reused from cycle to cycle.

◆ Conveyor units (two types): With power prerinse and without power prerinse.

Location and Set-Up

The distance water has to be piped from the water source to the dishwasher should be as short as possible to prevent heat loss. (See *Water Supply* later in this chapter.) The unit should be raised at least six inches off the floor for easy cleaning. The unit should include an easy-to-read water temperature thermometer. Near the machine, post instructions stating the water temperature and pressure settings and the amount and concentration of detergent.

Clean-in-Place Equipment

This equipment is designed to be cleaned by passing through it a detergent solution, hot-water rinse, and sanitizing solution. Examples include soft-serve ice cream and frozen yogurt dispensers.

◆ Cleaning and sanitizing solutions must remain in the pipes for a fixed time period.

◆ All food-contact surfaces must be cleaned.

◆ Washing, rinsing, and sanitizing liquids must not leak into the rest of the machine and must drain completely.

◆ If not designed to be taken apart for cleaning, the machine must include a system, such as a removable panel, to check if cleaning is complete.

Portable Equipment

This equipment makes cleaning easier and allows you to use a work space for more than one purpose. Choose equipment that one employee can roll or pick up.

Immobile Equipment

This equipment is sealed to a masonry base or is on legs raised six inches off the floor. Leave a one- to four-inch toe space.

Follow the manufacturer's instructions for leaving space from the wall or between other pieces of equipment. Immobile counter-top equipment should be on legs that allow a four-inch space or should be sealed to the counter with nontoxic food-grade sealant. The FDA allows less space for small equipment.

Cantilever Mounted Equipment

This equipment is attached to a wall bracket, leaving space behind and below the unit. Follow the manufacturer's instructions for mounting.

Utilities

Your water supply and electrical service should support your cleaning program and never endanger food.

Water Supply

Your water supply must be potable (drinkable) water under enough pressure to run cleaning equipment, such as dishwashers. If your supply is a private well, have it regularly checked by the health department. If you use bottled water, buy it from a supplier who complies with local codes. Bottled water must be dispensed from its original container.

Check the water heater's *recovery rate,* the length of time it takes to produce hot water once the supply is low enough to start refilling. Check the size of the holding tank. You need to have enough water whenever you run the dishwasher or fill dishwashing sinks. Check the distance from the water heater to the dishwashing machine or sinks. A booster heater may be needed to reach 180°F (82.2°C) in a high-temperature dishwashing machine. The pressure for the final sanitizing rinse should be between 15 pounds per square inch (psi) (100 kilopascals [kPa]) and 25psi (170kPa). Check the pump's temperature and pressure gauge that leads directly into the hot water rinse line.

Plumbing

Bad plumbing can contaminate water. Cross-connections and sewage are the major hazards.

Cross-Connection

A *cross-connection* is a link between your drinkable water system and unsafe water or chemicals through which backflow can occur. *Backflow* is the flow of water or other liquids or substances from a source of potential contamination into the drinkable water system. For example, if a faucet or hose is set too low in a sink, *back siphonage*, a type of backflow, can occur when the pressure in the safe water supply drops below the pressure of the unsafe water, sucking unsafe water into safe water.

To avoid cross-connection, backflow, and back siphonage:

◆ Set up an air gap at all outlets. An *air gap* is a clear air space that is between an outlet of drinkable water and the drain or the distance between the outlet of drinkable water and the highest possible water level or flood rim (see *Exhibits 9.8 and 9.9*). An air gap is the only completely reliable method to prevent backflow.

◆ Attach hoses only to sinks equipped with backflow preventive devices.

◆ Install backflow preventive devices between all connections in piping systems. All devices must meet American Society of Sanitary Engineers standards. Check local codes for requirements and work with a licensed plumber.

◆ Be extra careful with all food machines that are hooked up to the water supply. For example, the FDA requires carbonated drink machines without air gaps to have special backflow preventers, vents, or mesh screens between the carbonation device and the water supply line.

◆ Check all overhead waste water drain lines and sprinkler systems. Fix any water condensing on or dripping from pipes.

Sewage

Keep sewer water and solids from contaminating the food.

◆ All drains should be kept clear to prevent flooding.

◆ Any area where water is often spilled or that is cleaned by hosing down, such as a dishwashing area or kitchen, should have its own floor drain.

◆ All pipes containing nondrinkable water, such as those coming from toilets or sinks, should be identified by a plumber and labeled.

◆ Condensation drain outlets from equipment such as refrigerators and freezers should not directly touch or be attached to floor drains.

EXHIBIT 9.8 AIR GAP

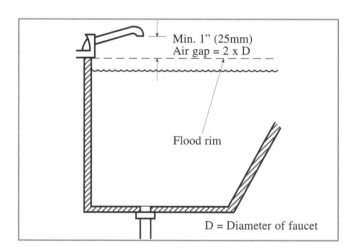

Electricity

Outlets and wiring should be safe and sufficient for your chilling, cleaning, and cooking equipment. Wiring and cords should not be frayed and plugs should not be cracked. Conduits should never run over the surface of interior walls.

EXHIBIT 9.9 SINKS WITH AIR GAPS

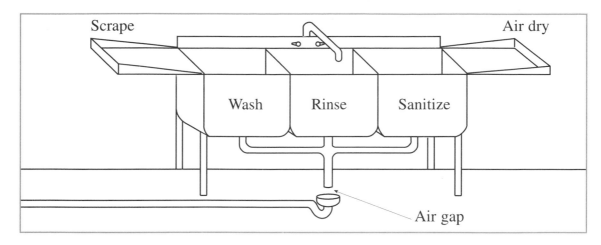

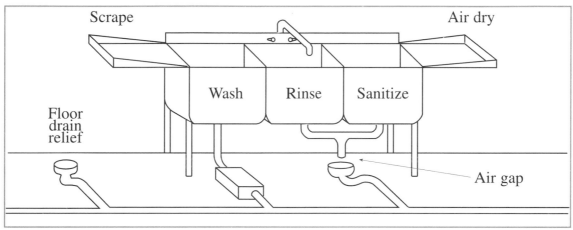

Courtesy of the Michigan Department of Public Health, Lansing, Michigan

Lighting

Lighting should be:

◆ Bright enough to reveal dirt and stains.

◆ Set so employees do not cast shadows on their work.

◆ Placed so broken glass cannot fall into food or other supplies. If lights must be placed near food, use shatter-proof bulbs and cover the lights.

• Cover incandescent lights with plastic panels, metal covers, or unbreakable glass globes.

• Cover fluorescent tubes with plastic sleeves that have endcaps.

• Cover heat lamps with shields.

Ventilation

Ventilation removes steam, smoke, grease, and heat from equipment and food preparation areas then takes makeup air (clean air) and replaces the air that was removed. Ventilation is necessary for reducing odors, gases, fumes, dirt, humidity, and grease build-up.

- ◆ Check local laws if the air must be cleaned before vented outside the building.

- ◆ Do not open windows or doors, unless they are screened, to ventilate areas.

- ◆ Use exhaust hoods over cooking, frying, and grilling equipment, steam tables, and dish- and pot-washers.

- ◆ Use hood filters or grease traps that are tight-fitting and easily removable and cleanable. Clean them often.

- ◆ Use screened outside air intakes to keep out pests.

- ◆ Avoid drafts across food preparation areas.

Garbage Disposal

Garbage is wet waste, usually from food. It can attract pests and be a source of contamination. Garbage disposal begins with keeping garbage away from food and food-contact surfaces.

- ◆ Remove garbage as soon as possible. Garbage from other areas should not be carried through food preparation areas.

- ◆ Put garbage in containers that are durable, leak-proof, easily cleanable, and pest- and water-proof. Containers may be metal or plastic and may be lined with plastic or wet-strength paper bags. Outside containers must have tight-fitting lids.

- ◆ Provide enough containers and dumpsters to hold all garbage between pick-ups. Store the containers on or above smooth surfaces that repel liquids, such as sealed concrete, in a cleanable, pest-free area away from food storage and preparation areas.

- ◆ Regularly clean and sanitize containers. Use an area away from food storage and preparation areas and equipped with hot and cold water and a floor drain.

EXHIBIT 9.10 GARBAGE DISPOSAL

Solid Waste Management

Solid waste includes dry, bulky trash, such as glass bottles, plastic wrappers and containers, paper bags, and cardboard boxes.

◆ Use pulpers or grinders to cut solid waste into small pieces that are flushed away with water. The water is removed and the solid waste is taken away.

◆ Use mechanical compactors to compress cans or cartons. This process requires a strong power source in a cleanable area with a drain.

◆ Practice *source reduction*, decreasing the amount of material received and disposed.

◆ Recycle items, such as paper, cardboard, polystyrene, glass, aluminum, tin, and used cooking oil. Check out local programs and local laws for storing and hauling recyclables.

◆ Join local waste-to-energy incineration programs, where a municipality burns waste to produce power.

A Case in Point

In 1986, several people became ill shortly after having iced drinks at a local restaurant with a bar lounge. They all complained that their iced drinks tasted like chemicals. The lounge's glassware washing machine was out of service.

When questioned, the manager said that the machine's wash cycle was working but that the unit could not be used because the large amount of water released after each wash load was making a drain blockage problem worse. A maintenance worker mentioned that the plumbing had backed up during the last week and that a large pool of water had been found under the glassware washing machine. Drain cleaners had been used repeatedly without any improvement.

The ice-making machine shared piping with both the glassware washing machine in the bar and the grease trap on the sink in the restaurant. The manager revealed that he himself had installed the plumbing fixtures. When the ice cubes were removed from the ice bin, thickened food grease and typical washing waste were found in the bin. Garbage also covered the bottom of the ice bin where the water drained out of the unit and into the pipes.

Why do you think the people became ill?

What should the manager do to correct this problem?

Answer to a Case in Point

The people who had drinks with ice became ill from the chemical drain cleaner that was used in the glassware washing machine. The pipe connection to the ice-making machine lacked an air gap to prevent backflow. Remember that the grease trap in the restaurant sink and the glassware washing machine drain were both connected to the same pipe as the ice-making machine. There were no air gaps or protective devices. As a result, waste and drain cleaner backed up into the ice cube bin, contaminating the ice used for drinks.

Ice bins and drainlines used for dispensing drink machines must be protected by approved air gaps. The manager needs to consult a licensed plumber to correct his plumbing system to meet local regulations.

Chapter 9 Exercise

1. Cleanable facilities and equipment are those from which soil and waste can be regularly removed by:

    a. hired professional cleaners.

    b. normal cleaning methods.

    c. closing for two days.

    d. only using strong chemicals.

2. Floors in kitchen, storage, and work areas should be:

    a. carpeted to reduce noise.

    b. nonskid, able to withstand strong cleaners, and should repel liquids.

    c. washed only with cold water and mild detergent.

    d. lightly cleaned once a week.

3. The safest materials for cutting boards are:

   a. food-grade seamless hard rubber and acrylics.
   b. soft wood, such as pine and balsa.
   c. stainless steel.
   d. rigid types of cardboard.

4. Which one of the following is a proper practice for clean-in-place equipment?

   a. Cleaning solutions must run straight through the pipes without stopping.
   b. Only sanitizing liquids can leak into the machine.
   c. Cleaning solutions must remain in pipes for a fixed time period.
   d. Only cold water should be used with sanitizing solutions.

5. Surfaces in refrigerators and freezers should be:

   a. kept damp to prevent dirt from sticking.
   b. able to absorb excess moisture.
   c. nontoxic and not leave a color, odor, or taste with food.
   d. painted with a lead-based paint.

6. Which one of the following should you do if your water heater does not heat the water to 180°F (82.2°C) or higher for your high-temperature dishwashing machine?

   a. Use luke-warm water in the wash cycle.
   b. Only wash small loads.
   c. Use extremely strong detergents and sanitizers.
   d. Install a booster heater.

7. A cross-connection is the:

   a. proper way to link two water systems.
   b. link between your water system and unsafe water.
   c. piping used in most dishwashers.
   d. piping used in drink dispensers.

8. An air gap is the:

   a. clear air space between an outlet of drinkable water and the drain.
   b. distance from a dishwasher sprayer to the dishes.
   c. width of a faucet opening.
   d. space under a piece of equipment and away from the wall.

9. Which one of the following is a good ventilation practice?

   a. Open windows or doors.
   b. Exhaust hoods over cooking equipment and dishwashers.
   c. Loose-fitting, easily cleanable air filters.
   d. Air intakes close to exhaust fans.

10. If you have more garbage than what fits in your containers, you should:

   a. carefully pile it next to and on top of the full containers.
   b. put it in the dry storage area.
   c. buy more containers at once.
   d. store it in the walk-in.

CHAPTER 10: CLEANING AND SANITIZING

Test Your Food Safety IQ

1. **True or False:** In general, the hotter the water, the better the detergent works. (See *Cleaning*, page 131.)

2. **True or False:** Hand soap is still the best way to clean most surfaces. (See *Cleaning Agents*, page 131.)

3. **True or False:** For all sanitizers, rinsing off detergent is no longer necessary. (See *Chlorine and Iodine*, page 133.)

4. **True or False:** Use test kits to regularly test sanitizer strength. (See *Chemical Sanitizing*, page 134.)

5. **True or False:** All bleaches can be used as sanitizers. (See *Scented or Oxygen Bleaches*, page 133.)

Learning Objectives

After completing this chapter, you should be able to:

◆ Supervise cleaning and sanitizing throughout your operation.

◆ Ensure safe machine and manual warewashing.

◆ Provide safe storage for clean and sanitized items.

◆ Train employees to safely handle cleaning supplies, including hazardous materials.

◆ Organize, implement, and monitor a cleaning program.

Cleaning and Sanitizing

All *food-contact surfaces* must be washed, rinsed, and sanitized:

◆ After each use.

◆ When you begin working with another type of food.

◆ Any time you are interrupted during a task and the tools or items you are working with may have been contaminated.

◆ At four-hour intervals if the items are in constant use.

Most other items should be cleaned and sanitized at least once each day and whenever they become soiled. This is true for food-contact items like grill surfaces and griddles and for non-food contact surfaces, such as brushes, mops, and buckets.

Cleaning

Several factors influence cleaning (see *Exhibit 10.1*).

EXHIBIT 10.1 FACTORS THAT INFLUENCE CLEANING

SOIL TYPE	Protein-based (blood, egg); grease or oil (margarine, animal fat); water miscible (dissolved in water—flour, starches, drink stains); acid or alkaline (tea, dust, wine, fruit juice)
SOIL CONDITION	Fresh, soft, ground-in, dried, or baked-on
WATER HARDNESS	Amount of dissolved minerals (calcium, magnesium, iron) in water. Measured in parts per million (ppm), grains per gallon, or milligrams per liter (mg/l)
WATER TEMPERATURE	Hotter the water, the quicker the detergent dissolves and the better it cleans
SURFACE BEING CLEANED	Different surfaces need different cleaners and methods
TYPE OF CLEANING AGENT	Appropriate for the item
AGITATION OR PRESSURE	Agitation, or scouring action, often needed to loosen soil
LENGTH OF TREATMENT	Longer the cleaning agent touches the surface, the better the cleaning

---

Cleaning Agents

Cleaning agents are chemical compounds made to remove soil or mineral deposits. In general, cleaning agents must work well, be stable, noncorrosive, and safe for employees to use.

Detergents

All *detergents* contain *surfactants*, substances that lessen surface tension between the detergent and the soiled surface so the detergent can penetrate and loosen soil. Most detergents also use alkaline substances to break up soil. Mild alkaline detergents are used to remove fresh soil from walls, floors, ceilings, and most equipment and utensils. Strong alkaline detergents are used to cut through wax, grease, and aged, baked, or burnt-on soil.

Solvent Cleaners

Solvent cleaners, often called *de-greasers*, are alkaline detergents that include a grease-dissolving agent. These cleaners work well on grill backsplashes, oven surfaces, and even grease stains on driveways. Solvent cleaners lose strength when diluted and are too costly to be regularly used on large areas.

Acid Cleaners

Acid cleaners are used when regular alkaline cleaners do not work. For example, they are used for scaling in dishwashing machines, rust stains in restrooms, and tarnish on copper and brass. They must always be used carefully and according to the manufacturer's instructions.

Abrasive Cleaners

Abrasive cleaners contain scouring agents that can be rubbed or scrubbed on hard-to-remove soils. These cleaners are often used on floors or baked- and burnt-on soils in pans. Abrasives may make cleaning harder and may scratch surfaces, such as Plexiglas, plastic, and even stainless steel.

Sanitizing

Sanitizing means reducing the harmful micro-organisms on a surface to safe levels. It is not a substitute for cleaning—food-contact surfaces must be cleaned and rinsed before they can be effectively sanitized.

Heat Sanitizing (see *Exhibit 10.4*)

In heat sanitizing, the temperature of food-contact surfaces must be 165°F (73.9°C) to kill micro-organisms. Check the temperature of the water and the temperature of the items you sanitize. Two methods for checking item temperatures are labels that change color when the required temperature is reached and paraffin or wax tape that melts at the required temperature.

Chemical Sanitizing

Chemical sanitizing solutions are widely used in the foodservice industry because of their effectiveness, reasonable cost, and easy use. These sanitizers are regulated by federal and state Environmental Protection Agencies (EPAs), which classify them in the same category as pesticides. Labels must state concentrations, effectiveness, directions for use, and possible health hazards.

Chlorine and Iodine

Items must be rinsed carefully before sanitizing. If used correctly, both chemicals are relatively easy on skin (see *Exhibits 10.2* and *10.3*). Chlorine compounds are more likely to damage rubber and metals, such as pewter, stainless steel, aluminum, and silverplate. Using too much of a chlorine-based solution may leave an odor on dishes. Iodine compounds should be used only in solutions with a pH of 5.0 or less, unless the manufacturer allows a higher limit.

Quaternary Ammonium Compounds (Quats)

Quats can work well in both acid and alkaline solutions and, if used correctly, are fairly easy on skin (see *Exhibits 10.2* and *10.3*). Quats may not work with all soaps and detergents. Mineral deposits in hard water may make some bacteria harder for quats to remove. Use quats only in water with a hardness of 500 ppm or lower. Check local regulations to see that the chemical makeup of your solution will work with the water in your area.

Change the solution when it becomes soiled, leaves the proper temperature range, or falls below the required concentration.

Other types of chemical sanitizers include:

◆ *Detergent-Sanitizer Blends:* These blends can be used to sanitize, but items must still be cleaned, rinsed, and then sanitized. These compounds are usually more expensive and may not be meant for food-contact surfaces.

◆ *Scented or Oxygen Bleaches:* These bleaches are not acceptable as sanitizers for food-contact surfaces and may leave residues. Household bleaches are acceptable if the labels indicate that they are EPA-registered.

EXHIBIT 10.2 FACTORS THAT INFLUENCE CHEMICAL SANITIZERS

CONTACT	Solution must contact items for the time specified by local regulations—usually one minute.
SELECTIVITY	Quats, in particular, may not kill all types of micro-organisms.
CONCENTRATION	Concentrations below the legal minimum may not sanitize. Concentrations above a certain level may leave a taste or odor, corrode metals, or qualify as a health department violation. It is best to use a system that automatically properly dilutes sanitizers.
TEMPERATURE	75° to 120°F (23.9° to 48.9°C). Solutions at the lower end of this range last longer. Below this range, chlorine and iodine do not work as well. Above this range, chlorine may corrode metals and both chlorine and iodine may evaporate.

EXHIBIT 10.3 GENERAL GUIDELINES FOR CHEMICAL SANITIZERS

	CHLORINE	IODINE	QUATERNARY AMMONIUM
Minimum Concentration — For Immersion	50 parts per million (ppm)	12.5—25.0 ppm	220 ppm*
— For Power Spray or Cleaning or Cleaning in Place	50 ppm	12.5—25.0 ppm	220 ppm*
Temperature of Solution	75°F (23.9°C)+	75°—120°F (23.9°—48.9°C) Iodine will leave solution at 120°F (48.9°C)	75°F (23.9°C)+
Time for Sanitizing —For Immersion —For Power Spray or Cleaning in Place	1 minute Follow manufacturer's instructions	1 minute Follow manufacturer's instructions	1 minute; however, some products require longer contact time—read label
pH (detergent residue raises pH of solution so rinse thoroughly first)	Must be below 8.0	Must be below 5.0	Most effective around 7.0 but varies with compound
Corrosiveness	Corrosive to some substances	Noncorrosive	Noncorrosive
Response to Organic Contaminants in Water	Quickly inactivated	Made less effective	Not easily affected
Response to Hard Water	Not affected	Not affected	Some compounds inactivated but varies with formulation—read label; hardness over 500 ppm is undesirable for some quats
Indication of Solution	Test kit required	Amber color indicates effective solution, but test kits must also be used	Test kit required; Closely follow label instructions

* Varies based on manufacturer's compounds.

Machine Cleaning and Sanitizing

High-temperature and chemical sanitizing dishwashing machines can help your operation handle a high volume of washing.

General Procedures

- ◆ Check the cleanliness of and clean each machine as often as needed, at least daily. Wash and rinse tanks should fill with clear water. Detergent trays and nozzles on spray arms should be clear.

- ◆ Flush, scrape, or soak items before washing. Pre-soak items with dried-on food.

- ◆ Correctly load the dishwasher racks—*never* overload them. This increases efficiency and helps ensure one-pass washing. Use racks constructed to expose all surfaces to the cleaning solution.

- ◆ Check temperatures.

- ◆ Check all items as they are removed. Run soiled dishes through again. Proper equipment and procedures will help ensure one-pass washing.

- ◆ Air dry all items. Do not use towels.

- ◆ Keep machines in good repair.

High-Temperature Machines

These machines rely on hot water to clean and sanitize (see *Exhibit 10.4*). You may need to add a booster heater to your water system to provide enough hot water to reach required temperatures. Each machine also needs a built-in thermometer to measure temperature at the manifold, where the water sprays into the cleaning chamber.

Chemical Sanitizing Machines

Chemical sanitizing units generally require water temperatures from 120° to 140°F (48.9° to 60°C). Use the right concentration of the manufacturer-recommended chemical. Use a machine in which the chemical sanitizing solution is automatically dispensed into the final rinse water (see *Exhibit 10.4*).

EXHIBIT 10.4 PROPER CLEANING, RINSING, AND SANITIZING TEMPERATURES

TEMPERATURE	PROCEDURE
75°–120°F (23.9°–48.9°C)	Range for chemical sanitizers to be effective
110°–120°F (43.3°–48.9°C)	Wash and rinse water temperature for manual immersion of tableware and equipment
120°–140°F (48.9°–60.0°C)	Range for sanitizing (final) cycle in chemical sanitizing machines (Check dishwashing-machine manufacturers' specifications)
150°F (65.6°C)	Wash solution for single-tank, stationary rack, dual-temperature machines
150°F (65.6°C)	Wash solution for multi-tank, conveyor, multi-temperature machines
160°F (71.1°C)	Wash solution for single-tank, conveyor, dual-temperature machines
165°F (73.9°C)	Quickly kills most micro-organisms
165°F (73.9°C)	Wash solution for single-tank, stationary rack, single-temperature machines
170°F (76.7°C) for 30 seconds	Heat sanitizing for manual immersion (Food and Drug Administration [FDA] requires 171°F [77°C] or higher)
180°F (82.2°C)	Final rinse cycle for a high-temperature machine at the manifold
195°F (90.6°C)	Upper limit for heat sanitization by machine or manual process
200°F (93.3°C)	Heat sanitization using live, additive-free steam

Manual Cleaning and Sanitizing

If you do not use a dishwashing machine, set up a three-compartment sink away from food-preparation areas (see *Exhibit 10.5*). (Some health departments allow two-compartment systems; others require four compartments.) Include an area for scraping and prerinsing scraps into garbage containers or grinders. All sinks should have thermometers to measure water temperature. Provide a separate drain board for clean items.

Three-Compartment Sink

This method can be used on tableware, utensils, and detachable equipment parts.

1. Clean and sanitize all sinks and work surfaces before washing dishes.

2. Flush, scrape, or soak items before washing.

3. Wash items in the first sink in detergent solution at least 110°F (43.3°C). Use a brush or cloth to loosen the remaining soil.

4. Rinse in the second sink in clear water at least 120°F (48.9°C). Remove all traces of food and detergent.

5. Sanitize in the third sink, by submerging items in:

   - Hot water at least 170°F (76.7°C) for 30 seconds. (Some states require 180°F [82.2°C].) To prevent burns, train employees to use tongs, a rack, or basket to lower items into the water.

   Or in:

   - A chemical sanitizing solution at least 75°F (23.9°C) or follow the manufacturer's instructions. Test the solution with a test kit.

6. Air dry all items.

Wood surfaces, such as cutting boards, wood handles, and bakers' tables, need special care. Scour them in a detergent solution with a stiff-bristle nylon brush, rinse in clear water, and sanitize after each use. *Never* submerge wooden cutting boards in detergent or sanitizing solution.

EXHIBIT 10.5 THREE-COMPARTMENT SINK

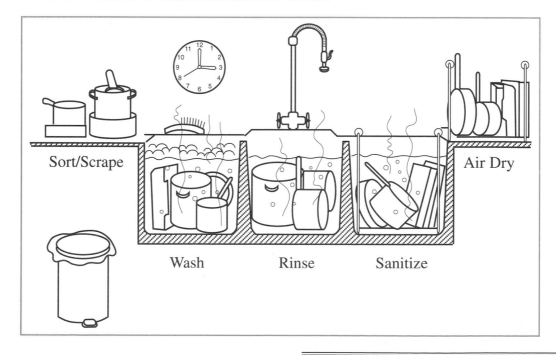

Sort/Scrape   Wash   Rinse   Sanitize   Air Dry

Cleaning Equipment

Several types of equipment need special cleaning and sanitizing procedures.

Clean-in-Place Equipment

This equipment is designed to have cleaning solution pumped through it. Follow the manufacturer's instructions. Cover food-contact parts when they are not in use. (See *Chapter 9*.)

Fixed or Immobile Equipment

Follow the manufacturer's instructions for cleaning. Food-contact surfaces usually require a different cleaning solution than nonfood-contact surfaces. Train employees to:

- ◆ Turn off and unplug equipment.

- ◆ Remove food and soil around and under the unit.

- ◆ Remove detachable parts and manually wash them. Turn the blades away from their bodies and wipe away from sharp edges.

- ◆ Wash, rinse, and sanitize the fixed food-contact surfaces. Wipe nonfood-contact parts with a sanitized cloth.

- ◆ Use marked buckets and separate cloths for food-contact and nonfood-contact surfaces.

- ◆ Air dry all parts.

- ◆ Put the unit back together and tighten all parts and guards. Plug in, set at recommended settings, and test. Turn settings back to "off" or zero.

- ◆ Resanitize food-contact surfaces that were handled when putting the unit back together.

Spray cleaning may be used if allowed by the manufacturer. Spray each part with the required concentration spray for two or three minutes. Steam cleaning may also be used on items where the steam can be held within the item. Steam should be at least 200°F (93.3°C). Both spray and steam cleaning should be done so that food and food-preparation areas are protected. Employees should be trained to use the equipment and wear proper protective gear.

Cooling and Microwave Units

The interiors of cooling and microwave units should be cleaned as often as necessary, at least daily, to remove spills, mold, or odors.

- Clean shelves, walls, floors, door edges and gaskets on all units.
- Clean cooling units before shipments arrive. Move stored food to another unit during cleaning.
- Clean microwave units according to manufacturer's instructions.

Cleaning the Restaurant

Clean walls, ceilings, floors, shelves, and light fixtures. Do not let cleaning solutions or water touch food or remain standing on floors or shelves.

Floors

Use the following steps for regular cleaning and for spills:

1. Set out cones or signs to mark the area being cleaned.
2. Sweep the area using dust-free compounds or vacuum cleaners.
3. Mop the area. Soak the mop in a bucket of detergent solution, squeeze out the mop, and wipe a 10-foot by 10-foot area, using both sides of the mop. Use scrub brushes for extra soiled areas, being careful not to splash. Mop away from walls and toward the floor drain.
4. Remove excess water with a mop or squeegee, again working away from walls and toward the floor drain.
5. Rinse thoroughly and again remove excess water.

Floor Drains

Clean drains as the last task of the day, after other cleaning is done.

1. Wear heavy rubber gloves and, if needed, rubber shoes or boots.
2. Remove the drain cover, remove waste, and replace the cover.
3. Flush the drain with a hose or spray, without splashing.
4. Pour cleaning detergent into the drain, scrub or spray the drain cover, and rinse.
5. Pour a sanitizing or disinfecting solution into the drain.

Ceilings

Ceilings do not need as much cleaning as floors, but should be checked daily for soil, cobwebs, and condensation. Wipe and rinse ceilings with a sponge.

Restrooms

Restrooms must be cleaned at least every day. Train employees to check for trash and spills every hour. Refill soap, toilet paper, and towel supplies before they are empty.

Storage

Store tableware, equipment, and cleaning supplies so they stay clean and sanitary.

Clean Tableware and Equipment

Train employees to:

◆ Clean and sanitize drawers and shelves before clean items are stored.

◆ Clean and sanitize trays and carts used to carry clean dishes from the storage area.

◆ Store tableware at least six inches off the floor and protected from soil and condensation.

◆ Store glasses and cups upside down. Store utensils so co-workers will grab them by the handles.

◆ Cover the food-contact parts of stored clean-in-place equipment.

Cleaning Supplies

Cleaning tools shall be cleaned and sanitized before being stored. They should be kept in a well-lighted, dry, locked area away from other chemicals, food, and items used to work on or prepare food. The storage area also needs a sink located near a floor drain to be used only for cleaning tasks.

Cloths, Sponges, and Scrubbing Pads

Store in a container of sanitizing solution or air dry.

Brushes and Mops

Store brushes hanging, rather than on their bristles. Store mops hanging, rather than standing in buckets.

Buckets and Pails

Store these items with other tools.

Use of Hazardous Materials

According to OSHA regulations, employees must be trained and supplied with the right equipment for handling hazardous materials.

◆ Inventory all hazardous chemicals.

◆ Be sure all substances containing hazardous chemicals are labeled. Labels should state the name of the chemical, the hazard, and the name and address of the manufacturer or responsible party.

◆ Get a *material safety data sheet* (MSDS) for each hazardous chemical. Store this information in a "right-to-know" station, where employees can easily reach it (see *Exhibit 10.6*).

◆ Be sure employees receive the following information (some of which is contained on the MSDS):

  • The common and chemical name of the product.

  • Where and when the product is to be used.

  • Proper training to use the product.

  • Physical hazards, such as fire, toxicity, or skin irritation.

  • Health hazards.

  • Emergency procedures to take if exposed to hazardous chemicals.

  • Protective steps for spills or leaks.

◆ Train employees to properly handle each chemical they work with and to protect themselves during use. Protective gear and emergency steps should be demonstrated.

Organizing a Cleaning Program

Each restaurant needs a cleaning program that is an overall system to organize all their cleaning and sanitizing tasks. Your program should help you identify your cleaning needs, set up a master cleaning schedule, select the supplies and tools you need, and train your employees to make the best use of their skills (see *Exhibit 10.7*).

EXHIBIT 10.6 SAMPLE MSDS

Courtesy of Ecolab®, St. Paul, Minnesotra

  OASIS™ 144 — **Quat Sanitizer** *Desinfectante Cuaternario*

```
985622              *MATERIAL SAFETY DATA SHEET*              Page 1 of 2
          MEDICAL EMERGENCY ONLY, 24 HOUR SERVICE: 1-800-328-0026

ECOLAB INC   Ecolab Center      Product Information: 1-612-293-2233
St. Paul MN 55102               Date of Issue: January 31, 1991
===============================================================
1.0 IDENTIFICATION /
1.1 Product Name:  OASIS 144
1.2 Product Type:  Quat Sanitizer

     ++ Section 2 Provides SARA Section 313 Reporting Information ++
-----------------------------------------------------------------
2.0 HAZARDOUS COMPONENTS /                    Air Limits (mg/m3)
                                              %   TWA   Other
2.1 Alkyldimethylbenzyl ammonium chlorides    10  None  UNK
              CAS 68424-85-1
     This product contains no other component considered hazardous
           according to the criteria of 29 CFR 1910.1200.
_____
3.0 PHYSICAL DATA /

3.1 Appearance:  Clear, dark purple liquid; sweet odor
3.2 Solubility in Water:  Complete
3.3 pH:  100% pH 7.5
3.4 Boiling Point:  > 212 deg F        Specific Gravity:  0.993
_____
4.0 FIRE AND EXPLOSION DATA /

4.1 Special Fire Hazards:  None
4.2 Fire Fighting Methods:  Product does not support combustion.
_____
5.0 REACTIVITY DATA /

5.1 Stability:  Stable under normal conditions of handling.
5.2 Conditions to Avoid:  Do not mix with anything but water.
_____
6.0 SPILL OR LEAK PROCEDURES /    USE PROPER PROTECTIVE EQUIPMENT

6.1 Cleanup:  Dike or dam large spills.  Pump to containers or soak up
     on inert absorbent.  Flush residue to sanitary sewer.
6.2 Waste Disposal:  Consult state and local authorities for
     restrictions on disposal of chemical waste.


     UNK = Unknown at this time    PEL = Permissible Exposure Limit
     TWA = OSHA 8 Hour Average     STEL = 15 Minute Average
     C = Ceiling Limit, Not To Be Exceeded
```

```
21495/0400/1090   Printed in U.S.A.   © Ecolab Inc. St. Paul, MN 55102
```

```
                                                    Page 2 of 2
      Center                                        985622
NLY, 24 HOUR SERVICE:  1-800-328-0026
=================================================
/                      DANGER

sure:
           cause severe irritation, possible chemical

ful.  Can cause chemical burns of mouth, throat

use irritation of mouth, throat and airways,
itive individuals.

ately with plenty of cool running water.  Remove
ntinue flushing for 15 minutes.
ith plenty of cool running water.  Wash
     thoroughly with soap and water.
8.3 If Swallowed:  Rinse mouth; then drink 1 or 2 large glasses of
     water.  DO NOT induce vomiting.  Never give anything by mouth to an
     unconscious person.
8.4 If Inhaled:  Move immediately to fresh air.

        CALL A POISON CONTROL CENTER OR PHYSICIAN IMMEDIATELY
_____
9.0 SPECIAL PROTECTION INFORMATION /

9.1 Respiratory:  Avoid breathing mists or vapors of this product.
9.2 Skin:  Rubber gloves - protective cuff or gauntlet type preferred.
9.3 Eyes:  Splashproof glasses, goggles or face shield.
_____
10.0 ADDITIONAL INFORMATION/PRECAUTIONS /

10.1 DOT:  Not DOT regulated.
10.2 This product is toxic to fish.  Do not discharge into lakes,
     streams, ponds or public waterways unless in accordance with an
     NPDES permit.  For guidance contact the regional office of the U.S.
     Environmental protection Agency.

            KEEP OUT OF REACH OF CHILDREN

The above information is believed to be correct with respect to the
formula used to manufacture the product.  As data, standards and
regulations change, and conditions of use and handling are beyond our
control, NO WARRANTY, EXPRESS OR IMPLIED, IS MADE AS TO THE
COMPLETENESS OR CONTINUING ACCURACY OF THIS INFORMATION.
```

EXHIBIT 10.7 SAMPLE MASTER CLEANING SCHEDULE

ITEM	WHAT	WHEN	USE	WHO
Floors	Wipe up spills	As soon as possible	Cloth, mop and bucket, broom and dustpan	
	Damp mop	Once per shift, between rushes	Mop, bucket	
	Scrub	Daily, closing	Brushes, squeegee, bucket, detergent (brand)	
	Strip, reseal	January, June	See procedure	
Walls and ceilings	Wipe up splashes	As soon as possible	Clean cloth, detergent (brand)	
	Wash walls	February, August		
Work tables	Clean and sanitize tops	Between uses and at the end of day	See cleaning procedure for each table	
	Empty, clean, and sanitize drawers; clean frame, shelf	Weekly, Saturday closing	See cleaning procedure for each table	
Hoods and filters	Empty grease traps	When necessary	Container for grease	
	Clean inside and outside	Daily, closing	See cleaning procedure	
	Clean filters	Weekly, Wednesday closing	Dishwashing machine	
Broilers	Empty drip pan, wipe down	When necessary	Container for grease, clean cloth	
	Clean grid tray inside, outside, and top	After each use	See cleaning procedure for each broiler	

A Case in Point

The student dishwashing employees in a university cafeteria had scraped every dish and utensil and placed all of them in the racks that automatically fed into the high-temperature dishwashing machine. When the wash and rinse cycles were complete, one of the employees noticed that the dishes were spotted and had bits of food on them. The thermometer that measured the final rinse temperature for the wash cycle read 100°F (37.8°C), rather than the required 140°F (60°C).

The employee informed the manager. The manager called the manufacturer's representative to come and examine the dishwashing machine.

What other methods can be used to clean and sanitize the tableware until the machine is fixed?

Answer to a Case in Point

The tableware can be washed, rinsed, and sanitized in a three-compartment sink using hot water and a chemical sanitizer. First, all three sinks must be cleaned and sanitized. Then, the tableware can be pre-soaked and washed in the first compartment and thoroughly rinsed in the second. The temperature in the wash compartment must be at least 110°F (43.3°C) and the rinse compartment must be 120°F (48.9°C). Chemical sanitizing can be done in the third compartment. The correct concentration should be used at a temperature between 75° and 120°F (23.9° and 48.9°C).

Tableware and equipment should always be air-dried and should not be handled until ready for use. Towel drying tableware can remove the chemical sanitizer before it has fully worked to kill micro-organisms, and unsanitary towels and hands can directly contaminate clean and sanitary dishes and utensils.

Chapter 10 Exercise

1. Factors that do *not* influence cleaning include:

   a. soil condition.
   b. soil texture.
   c. water temperature.
   d. water hardness.

2. Sanitize items with a chemical sanitizing compound or by:

   a. immersing them for 30 seconds in water that is at least 100°F (37.8°C).
   b. air drying them after washing.
   c. rubbing them with a towel.
   d. immersing them for 30 seconds in water that is least at 170°F (76.7°C).

3. Chemical sanitizers are most effective at temperatures from:

    a. 50° to 75°F (10.0° to 23.9°C).

    b. 75° to 120°F (23.9° to 48.9°C).

    c. 140° to 170°F (60.0° to 76.7°C).

    d. 200° to 212°F (93.3° to 100.0°C).

4. Dry sanitized items by:

    a. leaving them near an open window.

    b. air drying them.

    c. rubbing them with a towel.

    d. leaving them near a heat vent.

5. In a three-compartment sink system for manual warewashing:

    a. all sinks should contain detergent.

    b. all sinks should be at 200°F (93.3°C).

    c. wash sinks should be at least 110°F (43.3°C).

    d. wash sinks should be at least 180°F (82.2°C).

6. Store cleaning supplies:

    a. in the food preparation area.

    b. outside or on the loading dock.

    c. in the dry storage area.

    d. away from food, food-contact items, and other chemicals.

7. Store cleaning cloths, sponges, and scrubbing pads:

    a. in sanitizing solution or air dry them.

    b. on the floor under shelves.

    c. hanging in the kitchen.

    d. in flatware drawers.

8. A master cleaning schedule should list:

    a. outside cleaning firms to contact.

    b. cleaning violations from local ordinances.

    c. the date of the next visit from the health department.

    d. cleaning tasks, responsibilities, dates, and methods.

9. When using a dishwashing machine, employees should:

    a. ignore items with dried-on food.

    b. heavily load the dishmachine racks.

    c. towel dry all items.

    d. put soiled dishes in again.

10. Employees who handle hazardous materials should:

    a. ignore spills or leaks.

    b. have a basic idea of how to apply the product.

    c. be trained in emergency procedures for exposure to the product.

    d. forget about wearing protective gear.

CHAPTER 11: DEVELOPING AN INTEGRATED PEST MANAGEMENT (IPM) PROGRAM

Test Your Food Safety IQ

1. **True or False:** Managers themselves should apply all pesticides. (See *Pesticides*, page 153.)

2. **True or False:** Refuse any shipment of supplies that contains cockroaches, cockroach egg cases, or mice. (See *General Preventive Practices*, page 147.)

3. **True or False:** Rats need a hole only the size of a quarter to enter a building. (See *Pipes*, page 149.)

4. **True or False:** Cockroaches harm food but do not carry diseases. (See *Cockroaches*, page 150.)

5. **True or False:** Pesticide use is regulated by federal, state, and local laws. (See *Pesticides*, page 153.)

Learning Objectives

After completing this chapter, you should be able to:

◆ Set up an integrated pest management (IPM) program.

◆ Use methods to keep pests out of the building and off the grounds.

◆ Select methods for detecting pests.

◆ Identify methods to control pests.

◆ Work with a pest control operator (PCO).

Integrated Pest Management (IPM)

Pests, such as insects and rodents, are serious hazards to foodservice operations. Pests damage food, supplies, and facilities, but their greatest threat is that they spread diseases, including foodborne illnesses.

An integrated pest management (IPM) program is a system designed to prevent pests from infesting your restaurant and to get rid of any pests that are present. In an IPM program, you work closely with a licensed and registered pest control operator (PCO) to safely use various up-to-date pest control methods. Three common-sense rules for developing an IPM program are:

1. Deny pests food, water, and shelter by following good sanitation and housekeeping practices.

2. Keep pests out of the foodservice operation by pest-proofing the building.

3. Work with a licensed and registered PCO.

CAUTION! A number of pest control methods, including pesticides, if not used correctly can be hazardous to humans. Hiring a licensed PCO to manage your program and apply pesticides is highly recommended. If you do carry out your own pest control procedures, always consult a licensed PCO and check local regulations governing pesticide use.

Preventing Pest Infestations

A regular cleaning and sanitation program, as described in *Chapter 10*, is your first line of defense against pests.

General Preventive Practices

Start with these measures to keep all types of pests out of your restaurant:

◆ Use reputable and reliable suppliers. Check all supplies before they enter your building. Refuse any shipment in which you find pests, such as cockroaches (their egg cases) or mice.

◆ Remove garbage quickly and properly. Keep garbage tightly covered so it does not attract pests.

◆ Store recyclables as far from your building as local ordinances allow.

◆ Properly store all food and supplies:

• At least six inches off the floor and six inches away from walls.

• At 50 percent or less humidity. Low humidity helps keep cockroach eggs from hatching. Use dehumidifiers and ventilation equipment that force air outside the building.

• After they are opened, refrigerating foods, such as cocoa, powdered milk, and nuts, that attract insects may be another control method.

• Use FIFO to disturb insect breeding cycles.

◆ Keep cleaning equipment dry.

◆ Thoroughly clean and sanitize your restaurant.

• Immediately clean up food and beverage spills, including crumbs and scraps.

• Train employees to keep the break area and locker room clean. Food and dirty clothes should not be kept in or under lockers.

• Keep toilets and restrooms cleaned and sanitized.

Building and Grounds Maintenance

Holes and cracks, especially in older buildings, allow pests to enter and hide. Maintaining, repairing, and remodeling your building can reduce pest control costs.

Doors, Windows, and Vents

All openings should close tightly, be kept shut except when in use, and checked as part of a regular cleaning schedule.

- ◆ Windows and Vents: Use screening at least 16 mesh per square inch.

- ◆ Doors:

  - • Install self-closing devices and door sweeps.

  - • Repair gaps and cracks in door frames and thresholds.

- ◆ When necessary, install *air curtains*, also called *air doors* or *fly fans*, that put out a steady stream of air that flying insects avoid (see *Exhibit 11.1*). These doors should be wired in place to avoid being turned off.

EXHIBIT 11.1 AIR DOOR

Courtesy of Berner International, New Castle, Pennsylvania

Pipes

Rats need a hole only the size of a quarter to enter a building; mice need a hole only the size of a dime. Use concrete or sheet metal to fill or cover all holes around or near pipes. Install screening over ventilation pipes on the roof.

Floors and Walls

Keep rodents from entering or moving around your building by making needed repairs:

◆ Seal all cracks in floors and walls. Use sealants recommended by your PCO, health department representative, or building contractor.

◆ Close off spaces around or under fixed equipment. Use sealants for openings less than one-half inch (13 mm) and concrete or other materials for larger cracks (see *Exhibit 11.2*).

◆ Paint a white stripe around the edge of storeroom floors six inches from the wall. The stripe will remind employees to stack supplies away from the walls and rodent hairs, tracks, or droppings will show up against the white mark.

◆ Cover basement drains with holed, hinged metal caps to keep out rats and to make cleaning easier.

EXHIBIT 11.2 PROPER SEALING AROUND PIPES

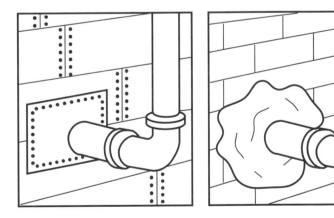

Indoor Garbage Storage Areas

To prevent odors and spills that attract pests:

◆ Keep garbage in sealed plastic bags inside tightly covered containers.

◆ Regularly wash, rinse, and sanitize the containers.

Grounds and Outdoor Serving Areas

To keep areas near the building pest-free:

◆ Mow the grass, get rid of standing water, and remove all litter and pet waste. Do not store building materials, old equipment, or any waste materials outside for any length of time.

◆ To avoid attracting bees, wasps, and other insects:

  • Cover all outdoor garbage containers and dumpsters.

  • Remove dirty dishes and uneaten food from tables and quickly clean up spills.

  • Avoid serving canned drinks outdoors. Bees and wasps often crawl inside the cans and may sting the drinker.

  • Allow only your PCO—not your employees—to remove hives and nests. These tasks take skill and are dangerous.

  • Keep insect zappers away from building doors, serving areas, food, employees, and customers. Use zappers that stun insects then trap them on glue paper.

◆ Do not allow employees or customers to feed birds on the grounds.

Spotting Signs of Pests

Even if you have a good pest control program, pests still may get into your building. Knowing how to spot them gives you a chance to contact your PCO and start early treatment.

Cockroaches

Roaches often carry disease-causing micro-organisms, such as *Salmonella*, fungi, parasite eggs, and viruses (see *Exhibit 11.3*).

Roaches may live and breed almost anywhere dark, warm, moist, and hard-to-clean:

◆ Behind refrigerators, freezers, and stoves.

◆ In sink and floor drains.

◆ In gaps around hot water pipes.

◆ In the motors of electrical equipment.

◆ Under shelf liners and wallpaper.

◆ Under rubber mats in beverage dispensers.

◆ In delivery boxes and bags.

◆ Behind walls and floor (especially rubber-based) coving.

Signs of Pests

Check for:

♦ A strong oily odor.

♦ Feces, which look like large grains of pepper.

♦ Egg cases which may be brown, dark brown, dark red, or black; capsule-shaped; and one-half inch (13 mm) long.

Note: A roach seen in daylight is usually a sign of a major infestation. Roaches normally search for food and water in the dark. Only the weakest members come out in the daylight.

Use glue traps to prove that you have roaches. *Glue traps* are cardboard containers open at both ends. Glue on the floor of the trap holds roaches that enter. Set traps where three surfaces meet, such as on the floor in a corner. Check the trap after 24 hours and show it to your PCO.

EXHIBIT 11.3 COMMON TYPES OF COCKROACHES

German	Pale brown or tan and ½ inch (13 mm) long. Found in indoor areas—in motors, crevices, soft-drink machines, and near water. Like potatoes, onions, and sweet beverages. Reproduce more rapidly than other roaches.
Oriental	Shiny, black, and about 1 to 1½ inches (25 to 38 mm) long. Found in basements, water pipes, and indoor incinerators.
American	Reddish brown and about 1½ inches (38 mm) long. Drawn to wallpaper, water, and starch in food. Found in drainage and sewer areas, restrooms, heating pipes, and damp open areas.
Brown-banded	Look like German roaches. Found in kitchens, dry storage, and under dining area tables and chairs.
Asian	Look like German roaches, but can fly. Are drawn to light and tend to hide in tropical plants.

Flies

The common housefly is an even greater threat to human health than the cockroach. Flies transmit several foodborne illnesses because they feed on garbage and animal wastes. They spread bacteria with their mouths, feet, hair, feces, and vomit. Flies have no teeth and eat only liquid or dissolved food. They vomit on solid food, let the food dissolve, and then eat it. Flies generally:

◆ Can enter a building through an opening the size of a pin head.

◆ Are drawn to odors of decay, garbage, and human and animal waste to lay their eggs and find food.

◆ Are drawn to places out of the wind and to the edges of objects, such as garbage can rims.

◆ Need moist, warm, decaying material protected from sunlight for their eggs to hatch into maggots. In warm summer weather, flies rapidly reproduce and maggots may grow into adult flies in six days.

Other Insect Pests

Beetles, moths, and ants can survive on very small amounts of food. Flour moth larvae, beetles, and such insects are found in dry storage areas. Signs of infestation include insect bodies, wings or webs, clumped-together food, and holes in food and packaging. Control these pests by tightly storing food in covered containers, using FIFO, and cleaning and sanitizing food preparation and serving areas.

Ants often nest in walls and floors, especially near stoves and hot water pipes. They are drawn to grease and sweet foods. Control ants by cleaning up all food scraps and spills.

Rodents

Rodents, also called vermin, eat and ruin food and damage property. They are a serious health hazard. They can spread disease through their waste and by touching food or food-contact surfaces. Rodents have a simple digestive system and weak bladder control. They urinate and defecate as they move about your facility, and their waste can fall or be blown or carried into food.

Rats have a highly developed sense of sound, touch, and smell. They can stretch themselves to reach 18 inches, climb vertical brick walls, and vertically jump three feet, horizontally four feet. Like other pests they breed quickly and often.

Signs of Infestation

Mice and rats can infest a building at the same time.

◆ *Droppings:* Fresh droppings are shiny and black. Older droppings are gray.

◆ *Gnawing:* Rats gnaw to reach food and to keep their teeth worn down. Rats can eat through pipes and hardened concrete as well as sacks, wood, and cardboard.

- *Tracks:* Check dusty surfaces by shining a light across them from a low angle.
- *Nesting Materials:* Mice use scraps of paper, hair, and other soft material.
- *Holes:* Rats nest in burrows, usually in dirt, rock piles, or along foundations.

Call your PCO if you think you have mice or rats. Rats are smart enough to avoid poorly set traps and other control measures. Rat bites are very dangerous to humans, and dead rats must be carefully handled to avoid spreading disease.

Birds

Common pigeons, starlings, and sparrows may nest in or near buildings. They eat animal and vegetable food items, and their droppings carry fungi that can make humans ill. Call your PCO to control birds. Your PCO has chemicals, electric shock equipment, netting, and traps to drive away birds.

Pesticides

Chemical pesticides are not a substitute for good sanitation. Do not try to solve pest problems by using pesticides without consulting your PCO. They can be dangerous to food, employees, and customers. They are toxic, hard to use, regulated by federal, state, and local laws, and can explode or cause fires. Aerosols can be improperly used and abused. Surfaces must be cleaned and sanitized after using aerosols.

Working with a PCO

Hire a licensed, certified, and reputable PCO to handle pest control procedures. You should work with a PCO because:

1. PCOs use methods that combine sanitation, nonchemical means, building maintenance, and chemical treatment.

2. PCOs know about new equipment and products.

3. Pest control is time-consuming. Your job is too complex and demanding to give adequate consideration to solving pest problems.

4. PCOs provide emergency service to help solve problems right away.

How to Hire a PCO

Carefully choose a PCO. Use these guidelines to make your decision:

1. Talk to restaurant managers about whom they use. When you call PCOs, ask for references and thoroughly check them.

2. Be sure the PCO is licensed or certified by the state, as required by federal law. Certification usually means that the PCO has passed an exam on pesticide use and other control procedures.

3. Ask whether the PCO belongs to professional organizations. Membership in a state or local association or the National Pest Control Association (NPCA) is a sign the PCO has up-to-date information.

4. Be sure the PCO has adequate insurance to cover the work—this protects you, your employees, and your customers. Ask for proof of insurance in the form of an insurance certificate.

Teamwork

Work with your PCO to arrange the best possible contract, inspection system, treatment procedures, and follow-up.

Contract

Always require a contract from your PCO. Have your lawyer review it. A contract should include:

◆ Warranty for work to be done and follow-up visits.

◆ Legal liability of the PCO.

◆ Length of service.

◆ Emergency service.

◆ Your duties.

◆ Records to be kept in a log book:

- Pests sighted or trapped: location, species, and actions taken.

- All chemicals used and accompanying MSDSs.

- Building problems noted and fixed.

- Maps or photos of trap and bait locations and problem areas.

- Schedule for trap, zapper, and chemical use.

- Reports from PCO.

Inspection

For an effective first inspection, you should walk around the building with the PCO and you should:

- Prepare employees to answer the PCO's questions.
- Have building plans and equipment diagrams available.
- Point out possible trouble spots.
- Have the PCO outline in writing the treatment plan:
  - Materials to be used.
  - Dates and times of treatment.
  - Steps you can take to control pests.
  - Building defects that may cause problems.

Treatment Procedures

Find out exactly which chemicals and procedures will be used in specific areas. Require the PCO to outline any risks of planned treatments, products, and equipment. Train employees to know and avoid any hazards.

Treatment Preparation

The federal government requires a PCO to give you sufficient advance warning to properly prepare your facility for the visit. Employees must be evacuated during the treatment.

Follow Up

Your pest control contract should mention follow-up visits after each treatment to review how well the treatment is working. You or your supervisors also should regularly check the restaurant and call the PCO if pests appear again.

PCO Methods

To control pests PCOs use several methods in ways that do not endanger employees and customers.

Controlling Insects

Repellents

Repellents are liquids, powders, and mists that keep insects away from an area but do not necessarily kill them. Repellents can be used in hard-to-reach places, such as behind wallboards and plaster. They are also used with glue traps, contact sprays, and residual sprays.

Sprays

Sprays are often used to control roaches and flies. It is easy to improperly use and abuse these. For safety, all foods and food-contact utensils must be removed from an area that will be sprayed. Immovable objects, such as counters and ovens, must be covered and then washed, rinsed, and sanitized after the area is sprayed.

◆ *Residual sprays* are used in cracks and crevices and leave a film of insecticide that insects absorb or pick up as they crawl across it.

◆ *Contact sprays* are used on groups of insects, such as clusters of roaches in a corner or a nest of ants. To be effective, the spray must come into *contact* with the insect.

Controlling Rodents

Rodents tend to reuse the same routes and runways. The PCO will choose the best method to clear the area.

Traps

Using traps is a slow but generally safe method of killing rodents. As bait, spring traps use food, such as peanut butter, that should be kept fresh. Traps should be set near rodent runways. They should be checked often and dead rodents should be carefully removed. If the trapped rodent is still alive, your PCO should remove it.

Glue Boards

Glue boards work for killing mice, but are usually not effective for killing rats. Boards do not contain poison and are safe to use. Mice get stuck to the board and die in several hours because of a lack of oxygen or water or from exhaustion. Boards should be put near mouse runways and checked often. Immediately throw out boards and mice.

Poison

Poisonous baits should only be used by a licensed PCO with extreme caution. Poison is added to the bait food and put near rodent runways. Poisoned baits should only be placed outdoors where they can not contaminate food or food-contact items. Bait should be set outside every day for at least two weeks. The PCO may change the type of bait and its location until the poisoned baits work. Employees must be careful to keep well away from these baits.

A Case in Point

The Now We're Cooking Restaurant is located in the middle of town on the first floor of a World War I landmark building. The building is in a shopping area that includes several other restaurants. Fred, the manager, recently had the interior of the restaurant remodeled with marble floors and new ceilings. He chose materials and equipment that are modern and easy to clean. The cleaning procedures for food preparation areas on Fred's master cleaning schedule are detailed, and his employees complete them as scheduled, with regular self-inspections. Spills and grease are cleaned up right away. Standard procedures include FIFO and using metal racks that keep food six inches off the floor and away from the walls.

During a self-inspection two weeks after the remodeling, Fred found cockroaches behind the sinks, in the vegetable storage area, in the public restrooms, and near the garbage storage area.

What factors might Fred have overlooked in his recent remodeling that could have led to this infestation?

What should Fred do to kill the cockroaches?

Answer to a Case in Point

Older buildings often have nooks and crannies that should be corrected with remodeling. Decaying outer building materials and less than sanitary neighbors can make pest control hard. Fred needs to have his contractor seal pipe openings and crevices around the sinks with a proper sealant. All cracks and crevices in the walls, floors, and door and window frames in the storage areas also should be sealed. It is important to clean the restrooms more thoroughly and more often. Any new procedures must be added to the master cleaning schedule and included in self-inspections.

Garbage should be taken out many times during the day. Garbage containers should be cleaned and sanitized after each load is emptied. Recyclables must be rinsed out and stored in a sanitary manner until they are picked up.

Fred should place glue traps in the problem areas and in other corners around the restaurant. He also should call a licensed and registered PCO. The PCO will be able to work with Fred to develop an IPM program to correct sanitation and structural flaws and to use nonchemical control methods, and, if necessary, safe and proper pesticides. As part of his IPM program, Fred should contact the businesses that share the building. They may all work with the PCO to improve their sanitation practices.

Chapter 11 Exercise

1. What level of humidity in storage areas helps control roaches?

   a. 120 percent.
   b. 100 percent.
   c. 75 percent.
   d. 50 percent.

2. What can you do to storage areas to see rodent signs better?

   a. Paint a white line on the floor six inches from the wall.
   b. Raise the humidity level.
   c. Leave the floor wet.
   d. Use a blue light bulb and dust the area.

3. Where should you install insect zappers?

   a. Near the customer entrances.
   b. By the dry storage area.
   c. Away from food and customers.
   d. Near buffet tables.

4. Seeing a cockroach in daylight usually means that you have:

   a. very few roaches.
   b. a large infestation.
   c. no rodents.
   d. low humidity.

5. To prove that you do have roaches, you should:

   a. use a glue trap.
   b. leave food on the counter.
   c. leave on the lights.
   d. spray pesticides near the refrigerator.

6. Which of the following is true of rodents?

   a. They leave waste everywhere they go.
   b. They are weak and easy to handle.
   c. They breed slowly and in small numbers.
   d. They cannot detect odors.

7. Putting out poisoned bait for rodents:

   a. is a simple, safe process.
   b. should be done only by you.
   c. should be done only by a licensed PCO.
   d. should be done only by experienced employees.

8. Which one of the following is true of pesticide use?

   a. Applying them often can take the place of good sanitation.
   b. They are flammable and nontoxic to humans.
   c. Their use is regulated by law.
   d. Anyone can apply them in a restaurant.

9. Which one of the following PCO methods keeps insects away but does not necessarily kill them?

   a. Residual sprays.
   b. Repellents.
   c. Glue traps.
   d. Contact sprays.

10. A contract with a PCO should include:

   a. every piece of equipment and building diagrams.
   b. recommendations from other restaurant owners.
   c. a list of professional organizations.
   d. a warranty for work to be done.

CHAPTER 12: REGULATORY AGENCIES AND INSPECTIONS

Test Your Food Safety IQ

1. **True or False:** Almost every aspect of a foodservice operation is regulated. (See *Government Regulation of the Food Industry*, page 160.)

2. **True or False:** Sanitation standards and their interpretation are the same nationwide. (See *Federal, State, and Local Regulatory Agencies*, page 161.)

3. **True or False:** State and local agencies have more control than federal agencies over a food service's day-to-day operation. (See *State and Local Regulatory Agencies*, page 161.)

4. **True or False:** Health department inspectors are often called sanitarians. (See *Restaurant Inspections*, page 162.)

5. **True or False:** The friendlier you are with the sanitarian, the better it is for your operation. (See *What to Do During an Inspection*, page 166.)

Learning Objectives

After completing this chapter, you should be able to:

◆ Understand the roles of federal, state, and local regulatory agencies.

◆ Prepare for various types of inspections.

◆ Work well with a sanitarian during an inspection.

◆ Record and make good use of inspection findings.

Government Regulation of the Food Industry

Almost every aspect of a foodservice operation is regulated by federal, state, and county or city agencies. These agencies have adopted standards that define sanitation regulations meant to protect the public against foodborne illnesses. Food processors, wholesalers, distributors, retailers, and restaurant operators must meet these standards.

It is up to each foodservice operator to learn the state and local codes and how they are enforced. Your strategy for doing this should include:

◆ Treating your state and local regulatory agencies as allies and asking them for help. Many state and local agencies provide training for managers.

◆ Meeting or going beyond safety requirements. Keeping your food safe improves its quality and makes your restaurant more cost-effective.

Types of Controls

◆ *Law:* The local, state, and federal statutes, regulations, and ordinances that apply in a jurisdiction.

◆ *Statute:* An established law that gives authority to make codes and regulations.

◆ *Code:* A systematic statement of a body of regulations that carries the force of law.

◆ *Regulation:* A rule that carries the force of law.

◆ *Ordinance:* A law set forth by a local authority, such as a municipality or county.

Federal, State, and Local Regulatory Agencies

Food and Drug Administration (FDA)

A number of federal agencies oversee food safety. Their regulations are published in the *Code of Federal Regulations* (CFR), which is sometimes cited in state and local codes. The most important agency among these is the Food and Drug Administration (FDA). (See the *Directory* at the end of the coursebook for information on other federal agencies.)

The FDA is part of the U.S. Department of Health and Human Services. Within the FDA, the United States Public Health Service develops model ordinances and codes for state and local health departments. Most importantly, the FDA publishes and updates the *Food Code* containing standards recommended for the entire food industry.

Although the FDA suggests adoption of its Food Code, it cannot require its adoption at the state or local level. Some state and local agencies develop their own codes.

State and Local Regulatory Agencies

State and local agencies have the most influence on the day-to-day operations of a food service. They are charged with publicizing and enforcing the local food code. Most importantly, these agencies conduct the regular sanitation inspections a restaurant must pass to stay in business. They may also issue sanitation guidelines that make the laws easier to understand. These guidelines are statements of recommended policies or procedures; they do not have the force of regulations or laws.

Restaurant Inspections

A *sanitarian*, sometimes called a *health official* or *inspector*, is trained in sanitation and public health. Sanitarians conduct plan reviews and pre-opening inspections for permit applications. They also carry out regular operational inspections and review applications for code variances.

Permit Applications, Plan Reviews, and Pre-Opening Inspections

Local agencies review applications for new operating permits and for permit renewals. Your local code may require that the agency reviews building plans for construction or remodeling. Once construction is completed, the sanitarian inspects the site before issuing an operating permit. The restaurant may be inspected again just before opening to be sure all food safety systems are in place. Several agencies, such as the fire department and OSHA, may be involved in the review.

Operational Inspections

Operational inspections are on-site visits by the local sanitarian to check if a restaurant's foodhandling practices meet the local codes (see *Exhibit 12.1* and *12.2*). The FDA recommends inspections at least every six months. Local agencies may schedule inspections monthly, quarterly, or annually based on several factors:

◆ The workload and number of employees in the agency.

◆ A history of violations or outbreaks.

◆ Serving many potentially hazardous foods, a large number of customers, or very young, elderly, or ill people.

Operations with good records or those with HACCP systems in place may need fewer on-site inspections.

EXHIBIT 12.1 SAMPLE INSPECTION FORM

Courtesy of DuPage County Health Department, Wheaton, Illinois

DuPAGE COUNTY HEALTH DEPARTMENT
ENVIRONMENTAL HEALTH DIVISION
FOOD SERVICE INSPECTION REPORT

Inspection Type _____

Establishment ID ☐☐☐☐☐

____ / ____ / ____
Date

NAME _____ ADDRESS _____

OWNER / OPERATOR _____ CITY _____

The items marked below identify violations of ORDINANCE No. 107-77. **CRITICAL items are to be corrected immediately.** All other items are to be corrected as soon as possible, but no later than the time specified on the subsequent page(s) of this report. Failure to comply may result in the suspension of your permit.

WT	X	SOURCE
5		1a. Approved source
5		1b. Wholesome, sound condition
1		2. Original container, properly labeled

TEMPERATURE CONTROL OF POTENTIALLY HAZARDOUS FOODS

WT	X	
5		3a. Cold food at proper temperatures during storage, display, service, transport, and cold holding
5		3b. Hot food at proper temperatures
5		3c. Foods properly cooked and/or reheated
5		3d. Foods properly cooled

FOOD TEMPERATURES (circled items are in violation)

_____ _____ _____
_____ _____ _____
_____ _____ _____

WT	X	
5		4. Facilities to maintain proper temperatures.
1		5. Thermometers provided and conspicuously placed
3		6. Potentially hazardous foods properly thawed

FOOD PROTECTION

WT	X	
5		7a. Cross-contamination, equipment, personnel, storage
1		7b. Potential for cross-contamination; storage practices; damaged food segregated
5		7c. Unwrapped food not re-served
3		8. Food protection during storage, preparation, display, service, transportation
3		9. Foods handled with minimum manual contact
1		10. In-use food dispensing utensils properly stored

PERSONNEL

WT	X	
5		11. Personnel with infections restricted
5		12a. Hands washed, good hygienic practices (observed)
1		12b. Proper hygienic practices, eating/drinking/smoking (evidence)
1		13. Clean clothes, hair restraints

FOOD EQUIPMENT AND UTENSILS

WT	X	
3		14. Food contact surfaces designed, constructed, maintained, installed, located
1		15. Non-food contact surfaces designed, constructed, maintained, installed, located
3		16. Dishwashing facilities designed, constructed, operated (1. wash 2. rinse 3. sanitize)
1		17. Thermometers, gauges, test kits provided
1		18. Pre-flushed, scraped, soaked
3		19. Wash, rinse water clean, proper temperature
5		20a. Sanitizing concentration _____ ppm
5		20b. Sanitizing temperature _____ °F
1		21. Wiping cloths clean, used properly, stored
3		22. Food contact surfaces of equipment and utensils clean
1		23. Non-food contact surfaces clean
1		24. Storage / handling of clean equipment, utensils

WT	X	SINGLE SERVICE ARTICLES
1		25. Single service items properly stored, handled, dispensed
3		26. Single service articles not re-used

WATER AND SEWERAGE / PLUMBING

WT	X	
5		27. Water source safe, hot and cold under pressure
5		28. Sewage and waste water disposed properly
1		29. Plumbing installed and maintained
5		30. Cross-connections, back-siphonage, back-flow prevented

HANDWASHING FACILITIES

WT	X	
5		31. Handwashing sinks installed, located, accessible
3		32. Restrooms with self-closing doors, fixtures operate properly, facility clean, supplied with handsoap, disposable towels or hand drying devices, tissue, covered waste receptacles

GARBAGE AND SOLID WASTE DISPOSAL

WT	X	
3		33. Containers covered, adequate number, insect and rodent proof, emptied at proper intervals, clean
1		34. Outside storage area clean, enclosure properly constructed

INSECT AND RODENT CONTROL

WT	X	
5		35a. Presence of insects / rodents. Animals prohibited
1		35b. Outer openings protected from insects, rodent proof

FLOORS, WALLS AND CEILINGS

WT	X	
1		36. Floors properly constructed, clean, drained, coved
1		37. Walls, ceilings, and attached equipment, constructed, clean
1		38. Lighting provided as required. Fixtures shielded
1		39. Rooms and equipment - vented as required

OTHER AREAS

WT	X	
1		40. Employee lockers provided and used, clean
5		41a. Toxic items properly stored
5		41b. Toxic items labeled and used properly
1		42. Premises maintained free of litter, unnecessary articles. Cleaning and maintenance equipment properly stored, kitchen restricted to authorized personnel
1		43. Complete separation from living/ sleeping area, laundry
1		44. Clean and soiled linen segregated and properly stored

Manager Certified? ☐ Y ☐ N ☐ N/A   Risk Type _____

Time In ___:___ ☐ am ☐ pm   Total Time _____

☐ Refer to page(s) _____ Comments: _____

Received By _____ Demerit Points _____ Follow-up _____

Sanitarian _____ Sanitarian ID __ __ __ __ Phone __ __ __ - __ __ __ __

AN OPPORTUNITY FOR APPEAL FROM ANY INSPECTION REPORTS WILL BE PROVIDED IF A WRITTEN REQUEST IS FILED WITH THE HEALTH AUTHORITY AS SPECIFIED IN ORDINANCE 107-77

REPORT MUST BE POSTED ON PREMISES

EXHIBIT 12.2 SAMPLE HACCP INSPECTION FORM

Courtesy of New York State Department of Health, Albany, New York

NEW YORK STATE DEPARTMENT OF HEALTH
Bureau of Community Sanitation and Food Protection

Hazard AnalysisCritical Control Point Monitoring Procedure Report

HACCP Page 1

COUNTY		DIST.			EST. NO.				MONTH		DAY		YEAR	

THIS FORM CONSISTS OF TWO PAGES AND BOTH MUST BE COMPLETED.

Establishment Name_____Operator's Name_____

Address_____

(T)(C)(V)_____County_____

Food_____

PROCESS (STEP) CIRCLE CCPs	CRITERIA FOR CONTROL	MONITORING PROCEDURE OR WHAT TO LOOK FOR	ACTIONS TO TAKE WHEN CRITERIA NOT MET
RECEIVING/ STORING	☐ Approved source (inspected) ☐ Shellfish tag ☐ Raw/Cooked/Separated in storage ☐ Refrigerate at less than or equal to 45°F	☐ Shellfish tags available ☐ Shellfish tags complete ☐ Measure food temperature ☐ No raw foods stored above cooked or ready to eat foods	☐ Discard food ☐ Return food ☐ Separate raw and cooked food ☐ Discard cooked food contaminated by raw food ☐ Food temperature: More than 45°F more than 2 hours, discard food More than 70°F, discard food
THAWING	☐ Under refrigeration ☐ Under running water less than 70°F ☐ Microwave ☐ Less than 3 lbs., cooked frozen ☐ More than 3 lbs., do not cook until thawed	Observe method Measure food temperature	Food temperature: More than or equal to 70°F, discard More than 45°F more than 2 hours, discard
PROCESSING PRIOR TO COOKING	Food temperature less than or equal to 45°F	Observe quantity of food at room temperature Observe time food held at room temperature	Food temperature: More than 45°F more than 2 hours, discard food More than 70°F, discard food
COOKING	Temperature to kill pathogens Food temperature at thickest part more than or equal to _____°F	Measure food temperature at thickest part	Continue cooking until food temperature at thickest part is more than or equal to _____°F
HOT HOLDING	Food temperature at thickest part more than or equal to _____°F	Measure food temperature at thickest part during hot holding every _____ minutes	Food temperature: 140°F - 120°F: More than or equal to 2 hours, discard; less than 2 hours, reheat to 165°F and hold at 140°F 120°F - 45°F: More than or equal to 2 hours, discard; less than 2 hours, reheat to 165°F and hold at 140°F

DOH-2614 (10/91) p. 1 of 2   White copy (pg. 1) - operator   Canary copy (pg. 1) - file

What Is Inspected

Many state and local ordinances are based on the FDA's *Food Code*. The main areas that this code covers include:

◆ Purchasing food from approved sources and in wholesome, sound condition.

◆ Observing the rules of time and temperature, controlling them throughout the flow of food.

◆ Providing facilities and equipment to maintain safe internal food product temperatures.

◆ Observing the rules for preventing cross-contamination.

◆ Restricting employees with infections who pose a hazard to food.

◆ Requiring strict personal hygiene of all employees.

◆ Installing hand sinks that are accessible to employees working in food preparation, warewashing, and service areas.

◆ Using proper sanitizer concentrations for all equipment, utensils, and food-contact surfaces.

◆ Providing a safe water supply and waste disposal system.

◆ Installing proper plumbing to prevent cross-connections, backflow, and back siphonage.

◆ Using measures to control and kill insects and rodents.

◆ Properly storing and labeling all cleaning agents and toxic materials.

Your best bet as a manager is to safeguard food everywhere in your restaurant. Get a copy of your local code and meet all its requirements.

Variances

For special reasons, a regulatory agency may allow a restaurant to vary from the local food code. *Variances* are changes or suspensions of the rules for a specific procedure or part of the restaurant. Agencies grant variances only when they are certain that the restaurant's practices are based on scientific principles and will not lead to a health hazard.

A restaurant must follow the procedures set up by the local agency and write to apply for each variance. The FDA recommends that applications include:

◆ A statement of the proposed change and how it differs from the existing code.

◆ An analysis of how the health hazards addressed by the existing code will continue to be prevented.

◆ Revisions to the restaurant's written HACCP plan (if a HACCP system is required by the local code) to cover the proposed variation.

If a variance is granted, the restaurant must comply with the new standards. Record keeping must be revised to document that the new procedures are safe. If a HACCP system is required, records should include CCPs, monitoring, corrective actions, and verification that the system is working.

What to Do During an Inspection

In most cases, the sanitarian will arrive for an inspection without warning. Local codes generally allow a sanitarian to inspect your records and establishment during operating hours or at any reasonable time.

You should ask the sanitarian for identification. Then find out if the visit is a regular inspection, the result of a customer complaint, or for another occasion. The FDA recommends that inspectors present their credentials and explain the purpose of their visit.

Take a positive, business-like approach when the sanitarian arrives for an inspection. Walk with the sanitarian during the inspection. This way you can immediately answer any questions and learn from the sanitarian's observations.

During the inspection, you should:

◆ Cooperate. Answer all questions as well as you can and instruct your employees to do the same.

◆ Take notes. Write down the sanitarian's comments, especially concerning any problems. Keeping an accurate record of the visit shows your concern and creates your own supply of data.

   If possible, correct problems on the spot. If you are sure the sanitarian is incorrect about something, take detailed notes on his or her comments. Review and discuss the specific code provision with the sanitarian. If necessary, request a written clarification from the director of the local regulatory agency.

◆ Keep the visit professional. Do not offer food or drink before, during, or after an inspection. Even these minor gestures can be seen as bribes.

◆ Be ready to provide records. The sanitarian may request documents, including employee files and records of all chemicals (such as for pest control) used on your premises. You should provide the requested records, but may ask why they are needed. The FDA recommends that local codes require that trade secrets, such as recipes, be kept secret.

◆ Closely study the inspection report. In detail discuss any violations. Record specific information to make the best possible corrections. Ask the sanitarian for advice on how to correct your problems.

After the inspection, the sanitarian will ask you to sign the inspection report. Most laws require signing it to acknowledge receipt. However, signing the report does not commit you to agreeing with or accepting the findings. At the same time, refusing to sign it does not free you from having to correct any violations. Regulatory jurisdictions will allow the chance for a hearing if you disagree with the report's findings.

◆ Follow up. Study your copy of the report:

- Decide why each violation occurred.

- Correct each violation.

- Complete corrections during the required time period. If you do not meet the deadline, you may be fined. The sanitarian may have the authority to close your operation if the violations are severe and are a "clear and present danger" to public health.

Finally, share the results of each inspection with your employees. Praise them for their efforts in the areas where they did well and work with them to improve problem areas.

A Case in Point

Carolyn, the registered sanitarian from the city's public health department, stood in the doorway of Jerry's Place. Jerry greeted her and they walked to the kitchen. Carolyn pulled out a HACCP worksheet. While Jerry explained that the cook was preparing a special of stir-fried chicken and vegetables, Carolyn noted the ingredients and their sources. Then she watched the preparation procedures and noted times and temperatures as she plotted them on a time/temperature graph. She also filled in a product flowchart and noted the CCPs for the chicken special. Jerry told her what corrective actions his employees were trained to carry out if the standards for the CCPs were not met.

Next, Carolyn checked the concentration of the sanitizing solution in the three-compartment sink that Jerry's dishwasher used for manually cleaning, rinsing, and sanitizing equipment. She also checked the foodhandlers' handwashing station.

Jerry and Carolyn went in Jerry's office, where they discussed the report. Jerry compared his flowchart for the stir-fried chicken and vegetables with the one Carolyn had made and determined where changes could be made to improve his monitoring system.

Did Jerry correctly handle the inspection?

What does Jerry need to do following this inspection?

Answer to a Case in Point

An inspection serves as a test of how well a food safety system is working. In this case, Jerry and his employees clearly have set up a good food safety system.

Jerry knew that stir-fried chicken and vegetables was a potentially hazardous menu item. He had designed a flowchart, setting up CCPs and corrective actions. He cooperated with Carolyn throughout the inspection and worked with her to improve his existing food safety system.

Jerry should continue doing his own self-inspections and monitoring his procedures to test how well the control measures are working. If corrective actions occur frequently, he may need to retrain some of his employees to follow the new procedures. He also may need to help them understand why the new system is better and how to make a commitment to the new procedures. Jerry knows that his food safety program is a day-to-day priority and not something to set up for inspections and then ignore.

Chapter 12 Exercise

1. Which one of the following must obey food safety laws?

   a. Food processors.
   b. Gardeners.
   c. Consumers.
   d. OSHA.

2. Many local food safety codes are based on the FDA's:

   a. Consumer Guide.
   b. Model Food Code.
   c. International Agreement.
   d. Dictionary of Foods.

3. Sanitarians:

   a. clean floors, counters, and dishes.
   b. manage quick-service stores.
   c. conduct inspections for the state or local health department.
   d. are trained to design thermometers and thermocouples.

4. Plan reviews are necessary when a restaurant:

   a. schedules new construction or remodeling.
   b. adds a new menu item.
   c. is inspected by a sanitarian.
   d. begins another year of business.

5. On-site visits by the local sanitarian to check if a restaurant's foodhandling practices meet the local code are called:

   a. hearings.
   b. one-on-ones.
   c. plan reviews.
   d. operational inspections.

6. The FDA recommends operational inspections at least every:

   a. week.

   b. month.

   c. six months.

   d. two years.

7. The main areas covered by the FDA *Food Code* include:

   a. time and temperature rules.

   b. sanitarian inspections.

   c. recipe secrets.

   d. proper advertising.

8. When the sanitarian arrives for an inspection:

   a. walk with him or her during the inspection.

   b. warn him or her that you expect fairness.

   c. avoid talking with him or her more than necessary.

   d. put him or her off until your work shift ends.

9. During the inspection, you should:

   a. offer the sanitarian food and drink.

   b. answer questions and provide the requested records.

   c. not allow employees to talk to the sanitarian.

   d. allow the sanitarian only one hour for the inspection.

10. If a restaurant does not meet the deadline set for correcting violations, a(n):

    a. restaurant may be fined or closed.

    b. extension is automatically granted.

    c. sanitarian will inspect the restaurant again the following year.

    d. restaurant cannot hire more employees.

Appendix A: Signs of Acceptable and Unacceptable Quality in Fresh Fruits

	Signs of Good Quality	Signs of Poor Quality, Spoilage
Apples	Firmness; crispness; bright color	Softness; bruises. (Irregularly shaped brown or tan areas do not usually affect quality)
Apricots	Bright, uniform color; plumpness	Dull color; shriveled appearance
Bananas	Firmness; brightness of color	Grayish or dull appearance (indicates exposure to cold and inability to ripen properly)
Blueberries	Dark blue color with silvery bloom	Moist berries
Cantaloupes (Muskmelons)	Stem should be gone; netting or veining should be coarse; skin should be yellow-gray or pale yellow	Bright yellow color; mold; large bruises
Cherries	Very dark color; plumpness	Dry stems; soft flesh; gray mold
Cranberries	Plumpness; firmness. Ripe cranberries should bounce	Leaky berries
Grapefruit	Should be heavy for its size	Soft areas; dull color
Grapes	Should be firmly attached to stems. Bright color and plumpness are good signs	Drying stems; leaking berries
Honeydew melon	Soft skin; faint aroma; yellowish white to creamy rind color	White or greenish color; bruises or watersoaked areas; cuts or punctures in rind
Lemons	Firmness; heaviness. Should have rich yellow color	Dull color; shriveled skin
Limes	Glossy skin; heavy weight	Dry skin; molds
Oranges	Firmness; heaviness; bright color	Dry skin; spongy texture; blue mold
Peaches	Slightly soft flesh	A pale tan spot (indicates beginning of decay); very hard or very soft flesh
Pears	Firmness	Dull skin; shriveling; spots on the sides
Pineapples	"Spike" at top should separate easily from flesh	Mold; large bruises; unpleasant odor; brown leaves
Plums	Fairly firm to slightly soft flesh	Leaking; brownish discoloration
Raspberries, Boysenberries	Stem caps should be absent; flesh should be plump and tender	Mushiness; wet spots on containers (signs of possible decay of berries)
Strawberries	Stem cap should be attached; berries should have rich red color	Gray mold; large uncolored areas
Tangerines	Bright orange or deep yellow color; loose skin	Punctured skin; mold
Watermelon	Smooth surface; creamy underside; bright red flesh	Stringy or mealy flesh (spoilage difficult to see on outside)

Source: HACCP Reference Book. Copyright © 1993 by The Educational Foundation of the National Restaurant Association.

Appendix B: Signs of Acceptable and Unacceptable Quality in Fresh Vegetables

	Signs of Good Quality	Signs of Poor Quality, Spoilage
Artichokes	Plumpness; green scales; clinging leaves	Brown scales; grayish-black discoloration; mold
Asparagus	Closed tips; round spears	Spread-out tips; spears with ridges; spears that are not round
Beans (snap)	Firm, crisp pods	Extensive discoloration; tough pods
Beets	Firmness; roundness; deep red color	Gray mold; wilting; flabbiness
Brussels sprouts	Bright color; tight-fitting leaves	Loose, yellow-green outer leaves; ragged leaves (may indicate worm damage)
Cabbage	Firmness; heaviness for size	Wilted or decayed outer leaves (Leaves should not separate easily from base.)
Carrots	Smoothness; firmness	Soft spots
Cauliflower	Clean, white curd; bright green leaves	Speckled curd; severe wilting; loose flower clusters
Celery	Firmness; crispness; smooth stems	Flabby leaves; brown-black interior discoloration
Cucumber	Green color; firmness	Yellowish color; softness
Eggplant	Uniform, dark purple color	Softness; irregular dark brown spots
Greens	Tender leaves free of blemishes	Yellow-green leaves; evidence of insect decay
Lettuce	Crisp leaves; bright color	Tip burn on edges of leaves (slight discoloration of outer leaves is not harmful)
Mushrooms	White, creamy, or tan color on tops of caps	Dark color on underside of cap; withering veil
Onions	Hardness; firmness; small necks; papery outer scales	Wet or soft necks
Onions (green)	Crisp, green tops; white portion two to three inches in length	Yellowing; wilting
Peppers (green)	Glossy appearance; dark green color	Thin walls; cuts, punctures
Potatoes	Firmness; relative smoothness	Green rot or mold; large cuts; sprouts
Radishes	Plumpness; roundness; red color	Yellowing of tops (sign of aging); softness

	Signs of Good Quality	Signs of Poor Quality, Spoilage
Squash (summer)	Glossy skin	Dull appearance; tough surface
Squash (winter)	Hard rind	Mold; softness
Sweet potatoes	Bright skins	Wetness; shriveling; shrunken and discolored areas on sides of potato (Sweet potatoes are extremely susceptible to decay.)
Tomatoes	Smoothness; redness. (Tomatoes that are pink or slightly green will ripen in a warm place.)	Bruises; deep cracks around the stem scar
Watercress	Crispness; bright green color	Yellowing, wilting, decaying of leaves

Source: HACCP Reference Book: *Copyright © 1993 by The Educational Foundation of the National Restaurant Association.*

Appendix C: Refrigerated Storage of Foods

FOOD	RECOMMENDED PRODUCT TEMPERATURES (°F/°C)	MAXIMUM STORAGE PERIODS	COMMENTS
Meat			
Roasts, steaks, chops	32–36/0–2.2	3 to 5 days	Wrap tightly
Ground and stewing	32–36/0–2.2	1 to 2 days	Wrap tightly
Variety meats	32–36/0–2.2	1 to 2 days	Wrap tightly
Whole ham	32–36/0–2.2	7 days	May wrap tightly
Half ham	32–36/0–2.2	3 to 5 days	May wrap tightly
Ham slices	32–36/0–2.2	3 to 5 days	May wrap tightly
Canned ham	32–36/0–2.2	1 year	Keep in can
Frankfurters	32–36/0–2.2	1 week	Original wrapping
Bacon	32–36/0–2.2	1 week	May wrap tightly
Luncheon meats	32–36/0–2.2	3 to 5 days	Wrap tightly when opened
Leftover Cooked Meats	32–36/0–2.2	1 to 2 days	Wrap or cover tightly
Gravy, Broth	32–36/0–2.2	1 to 2 days	Highly perishable
Poultry			
Whole chicken, turkey, duck, goose	32–36/0–2.2	1 to 2 days	Wrap loosely
Giblets	32–36/0–2.2	1 to 2 days	Wrap separate from bird
Stuffing	32–36/0–2.2	1 to 2 days	Covered container separate from bird
Cut-up cooked poultry	32–36/0–2.2	1 to 2 days	Cover
Fish			
Fatty fish	30–34/–1.1–1.1	1 to 2 days	Wrap loosely
Fish—not iced	30–34/–1.1–1.1	1 to 2 days	Wrap loosely
Fish—iced	32/0	3 days	Don't bruise with ice
Shellfish	30–34/–1.1–1.1	1 to 2 days	Covered container

FOOD	RECOMMENDED PRODUCT TEMPERATURES (°F/°C)	MAXIMUM STORAGE PERIODS	COMMENTS
Eggs			
Eggs in shell	40/4.4	1 week	Do not wash, or remove from container
Leftover yolks/whites	40–45/4.4–7.2	2 days	Cover yolks with water
Dried eggs	40–45/4.4–7.2	1 year	Cover tightly
Reconstituted eggs	40–45/4.4–7.2	1 week	Same treatment as eggs in shell
Cooked Dishes with Eggs, Meat, Milk, Fish, Poultry	32–36/0–2.2	Serve day prepared	Highly perishable
Cream-Filled Pastries	32–36/0–2.2	Serve day prepared	Highly perishable
Dairy Products			
Fluid milk	38–39/3.3–3.9	5 to 7 days after date on carton	Keep covered and in original container
Butter	38–40/3.3–4.4	2 weeks	Waxed cartons
Hard Cheese (cheddar, parmesan, romano)	38–40/3.3–4.4	6 months	Cover tightly to preserve moisture
Soft cheese			
Cottage cheese	38–40/3.3–4.4	3 days	Cover tightly
Other soft cheeses	38–40/3.3–4.4	7 days	Cover tightly
Evaporated milk	50–70/10–21.1	1 year unopened	Refrigerate after opening
Dry milk (nonfat)	50–70/10–21.1	1 year unopened	Refrigerate after opening
Reconstituted dry milk	38–40/3.3–4.4	1 week	Treat as fluid milk
Fruit			
Apples	40–45/4.4–7.2	2 weeks	Room temperature till ripe
Avocados	40–45/4.4–7.2	3 to 5 days	Room temperature till ripe
Bananas	40–45/4.4–7.2	3 to 5 days	Room temperature till ripe
Berries, Cherries	40–45/4.4–7.2	2 to 5 days	Do not wash before refrigerating
Citrus	40–45/4.4–7.2	1 month	Original container
Cranberries	40–45/4.4–7.2	1 week	
Grapes	40–45/4.4–7.2	3 to 5 days	Room temperature till ripe

FOOD	RECOMMENDED PRODUCT TEMPERATURES (°F/°C)	MAXIMUM STORAGE PERIODS	COMMENTS
Pears	40–45/4.4–7.2	3 to 5 days	Room temperature till ripe
Pineapples	40–45/4.4–7.2	3 to 5 days	Refrigerate (lightly covered) after cutting
Plums	40–45/4.4–7.2	1 week	Do not wash before refrigerating
Vegetables Sweet potatoes, mature onions, hard-rind squashes, rutabagas	60/15.6	1 to 2 weeks at room temperature; 3 months at 60°F (15.6°C)	Ventilated containers for onions
Potatoes	45–50/7.2–10	30 days	Ventilated containers
All other vegetables	40–45/4.4–7.2	5 days maximum for most; 2 weeks for cabbage, root vegetables	Unwashed for storage

Source: HACCP Reference Book: *Copyright © 1993 by The Educational Foundation of the National Restaurant Association.*

Appendix D: Storage of Frozen Foods

FOOD	MAXIMUM STORAGE PERIOD AT –10° TO 0°F (–23.3° TO –17.8°C)
Meat	
Beef, roasts and steaks	6 months
Beef, ground and stewing	3 to 4 months
Pork, roasts and chops	4 to 8 months
Pork, ground	1 to 3 months
Lamb, roasts and chops	6 to 8 months
Lamb, ground	3 to 5 months
Veal	8 to 12 months
Variety meats (liver, tongue)	3 to 4 months
Ham, frankfurters, bacon, luncheon meats	2 weeks (freezing not generally recommended)
Leftover cooked meats	2 to 3 months
Gravy, broth	2 to 3 months
Sandwiches with meat filling	1 to 2 months
Poultry	
Whole chicken, turkey, duck, goose	12 months
Giblets	3 months
Cut-up cooked poultry	4 months
Fish	
Fatty fish (mackerel, salmon)	3 months
Other fish	6 months
Shellfish	3 to 4 months
Ice Cream	3 months. Original container. Quality maintained better at 10°F (–12.2°C).
Fruit	8 to 12 months
Fruit Juice	8 to 12 months
Vegetables	8 months
French-Fried Potatoes	2 to 6 months
Precooked Combination Dishes	2 to 6 months
Baked Goods	
Cakes, prebaked	4 to 9 months
Cake batters	3 to 4 months
Fruit pies, baked or unbaked	3 to 4 months
Pie shells, baked or unbaked	1 1/2 to 2 months
Cookies	6 to 12 months
Yeast breads and rolls, prebaked	3 to 9 months
Yeast breads and rolls, dough	1 to 1 1/2 months

Source: HACCP Reference Book: *Copyright © 1993 by The Educational Foundation of the National Restaurant Association.*

Appendix E: Recommended Maximum Storage Periods for Goods in Dry Storage

FOOD	RECOMMENDED MAXIMUM STORAGE PERIOD IF UNOPENED
Baking Materials	
Baking powder	8 to 12 months
Baking soda	8 to 12 months
Chocolate, baking	6 to 12 months
Chocolate, sweetened	2 years
Cornstarch	2 to 3 years
Tapioca	1 year
Yeast, dry	18 months
Beverages	
Carbonated beverages	Indefinitely
Coffee, ground, not vacuum packed	2 weeks
Coffee, ground, vacuum packed	7 to 12 months
Coffee, instant	8 to 12 months
Tea, instant	8 to 12 months
Tea, leaves	12 to 18 months
Canned Goods	
Fruits (in general)	1 year
Fruits, acidic (citrus, berries, sour cherries)	6 to 12 months
Fruit juices	6 to 9 months
Pickled fish	4 months
Seafood (in general)	1 year
Soups	1 year
Vegetables (in general)	1 year
Vegetables, acidic (tomatoes, sauerkraut)	7 to 12 months
Dairy Food	
Cream, powdered	4 months
Milk, condensed	1 year
Milk, evaporated	1 year
Fats and Oils	
Mayonnaise	2 months
Salad dressings	2 months
Salad oil	6 to 9 months
Vegetable shortenings	2 to 4 months

FOOD	RECOMMENDED MAXIMUM STORAGE PERIOD IF UNOPENED

Grains and Grain Products

Cereal grains for cooked cereal	8 months
Cereals, ready-to-eat	6 months
Flour, bleached	9 to 12 months
Macaroni, spaghetti, and other noodles	3 months
Prepared mixes	6 months
Rice, brown or wild	Should be refrigerated
Rice, parboiled	9 to 12 months

Seasonings

Flavoring extracts	Indefinite
Monosodium glutamate	Indefinite
Mustard, prepared	2 to 6 months
Paprika, chili powder, cayenne	1 year
Salt	Indefinite
Sauces (steak, soy, etc.)	2 years
Seasoning salts	2 years
Spices and herbs (whole)	2 years to indefinite
Vinegar	1 year

Sweeteners

Sugar, brown	Should be refrigerated
Sugar, granulated	Indefinite
Sugar, confectioners	Indefinite
Syrups, corn, honey, molasses, sugar	1 year

Miscellaneous

Cookies, crackers	1 to 6 months
Dried beans	1 to 2 years
Dried fruits	6 to 8 months
Dried prunes	Should be refrigerated
Gelatin	2 to 3 years
Jams, jellies	1 year
Nuts	1 year
Pickles, relishes	1 year
Potato chips	1 month

Source: HACCP Reference Book: *Copyright © 1993 by The Educational Foundation of the National Restaurant Association.*

GLOSSARY

Acidity is measured on a pH scale from 0 (very acid) to 14.0 (very alkaline [basic]), with 7.0 being neutral. A pH level between 4.6 and 7.0 will support bacterial growth. (*Chapter 2*)

Air curtains, also called air doors or fly fans, are units that put out a steady stream of air that flying insects avoid. (*Chapter 11*)

Air gap is a clear air space that is between an outlet of drinkable water and the drain or the distance between the outlet of drinkable water and the highest possible water level or flood rim. (*Chapter 9*)

Aseptically packaged refers to food that has been hermetically sealed to prevent contamination from disease-causing micro-organisms. (*Chapters 6, 7*)

Back siphonage, a type of backflow, occurs when a loss of pressure in the water supply causes dirty water or chemicals to be sucked back into the drinkable water supply. (*Chapter 9*)

Backflow is a reversed flow of unsafe water into sinks and equipment, posing the threat of contamination to food, food-contact surfaces, and safe water supplies. (*Chapter 9*)

Bacterium is a living single-celled organism. (*Chapter 2*)

Biological hazard refers to the danger of food contamination by disease-causing micro-organisms (bacteria, viruses, parasites, or fungi), certain plants, and fish that carry toxins. (*Chapters 1,2*)

Carrier is a person or animal whose body harbors a disease-causing micro-organism. (*Chapter 2*)

Chemical hazard refers to the danger of food contamination by pesticides, food additives and preservatives, cleaning supplies, and toxic metals that leech through worn cookware and equipment. (*Chapters 1, 2*)

Ciguatera is a form of human poisoning caused by eating some tropical marine fish which through their diet have accumulated natural occurring toxins. (*Chapter 2*)

Clean means free of visible soil and food waste. (*Chapters 1, 9*)

Cleanable means that surfaces are accessible and soil and waste can be effectively removed by normal cleaning methods. (*Chapters 9, 10*)

Compensatory damages are awarded for lost work, lost wages, and medical bills the plaintiff may have experienced. (*Introduction*)

Contact spray is used on groups of insects. To be effective, the spray must come into *contact* with the insect. (*Chapter 11*)

Contamination is the unintended presence of harmful substances or disease-causing micro-organisms in food. (*Chapter 1*)

Coving is a curved, sealed ³/₈-inch edge between the wall and the floor. (*Chapter 9*)

Crash training is the attempt to cover a lot of material on the job in a very short time period. (*Chapter 4*)

Critical control point (CCP) is an operation (practice, preparation step, or procedure) where a preventive or control measure can be applied that would eliminate or prevent a hazard or lessen the risk that a hazard will happen. (*Chapter 4*)

Cross-connection is a link between a drinkable water system and unsafe water or chemicals. (*Chapter 9*)

Cross-contamination is the transfer of harmful substances or disease-causing micro-organisms to food by hands, food-contact surfaces, or cleaning cloths that touch raw food, are not cleaned and sanitized, and then touch ready-to-eat food. Cross-contamination can also occur when contaminated food or stored raw food touches or drips fluids on cooked or ready-to-eat food. (*Chapters 1, 3*)

Deep chilling is storing food at unit temperatures of 26° to 32°F (–3.3° to 0°C) for short time periods. (*Chapter 7*)

Dry lab is to record data without actually measuring the food's temperature. (*Chapter 4*)

FAT-TOM is an acronym for the conditions necessary for bacterial growth: food, acidity (pH), time, temperature, oxygen, and moisture (water activity). (*Chapter 2*)

First in, first out (FIFO) is a method of stock rotation, in which new supplies are shelved behind old supplies, so the old are used first. All inventory is marked with either the expiration date, when the item was received, or when it was stored after preparation. (*Chapter 7*)

Flowchart is a simple diagram that shows the flow of food and a recipe's critical control points (CCPs). (*Chapter 4*)

Flow of food is the path from receiving through storing, preparing, cooking, holding, serving, cooling, and reheating that foods follow in a foodservice operation. (*Chapters 1, 4*)

Foodborne illness is a disease that is carried or transmitted to people by food. (*Chapters 1, 2*)

Foodborne infection is a disease that results from eating food containing harmful micro-organisms. (*Chapter 2*)

Foodborne intoxication is a disease that results from eating food containing toxins from bacteria, molds, or certain plants or animals. (*Chapter 2*)

Food-contact surface is any equipment or utensil which normally comes in contact with food or which may drain, drip, or splash in food or on surfaces normally in contact with food. (*Chapters 1, 10*)

Fungi are a group of micro-organisms that include molds and yeasts. (*Chapter 2*)

Garbage is wet waste, usually from food. (*Chapter 9*)

Glue traps are cardboard containers open at both ends that are used to catch pests. The glue on the floor of the trap holds roaches or rodents that enter the trap. (*Chapter 11*)

Group training involves a group of trainees meeting with a trainer, usually in a session apart from their normal work. (*Chapter 4*)

Hazard Analysis Critical Control Point (HACCP) is a food safety system that focuses on the flow of food in a foodservice operation to reduce the risk of foodborne outbreaks. (*Chapters 4, 10*)

Hazards are items that may contaminate food at any time during its flow through a foodservice operation. Hazards include micro-organisms that can grow during preparation, storage, and/or holding; micro-organisms or toxins that can survive heating; chemicals and objects that can contaminate food or food-contact surfaces. (*Chapter 4*).

Histamine is an odorless, tasteless chemical that may be produced in some fish that are temperature abused. High levels of histamine may cause scombroid intoxication. Histamine is not destroyed by cooking. (*Chapter 2*)

Host is a person, animal, or plant on which another organism lives and feeds. (*Chapter 2*)

Individual training, also called one-on-one training, assigns one or two trainees to an experienced employee to learn the task. (*Chapter 4*)

Layout is the order of equipment, work areas, and furniture for dining and back-of-the-house areas. (*Chapter 9*)

Material safety data sheets (MSDSs), which are required by the Occupational Safety and Health Administration (OSHA), are written descriptions of the contents, hazards, and handling procedures for chemicals and products containing chemicals. (*Chapters 7, 10*)

Micro-organism is a small life-form, only seen through a microscope, that may cause a disease. (*Chapter 2*)

Modified atmosphere package (MAP) is a sealed package in which the oxygen has been reduced or replaced with other gases. This type of packaging extends the food's shelf life. (*Chapters 6, 7*)

Molds are fungi that can spoil food or may produce poisonous toxins. (*Chapter 2*)

Outbreak is an incident in which two or more people experience the same illness after eating the same food. Laboratory analysis must show that the food is the source of the illness. (*Chapter 1*)

Parasite is a micro-organism that needs a host to survive. (*Chapter 2*)

Pathogenic refers to bacteria that are infectious and disease-causing. (*Chapter 2*)

Physical hazard refers to the danger posed to food safety by foreign matter—such as dirt, hair, nails, staples, metal fragments, and broken glass and crockery—that accidentally get into food. (*Chapters 1, 2*)

Pooling is to crack several eggs into a bowl. This is not recommended unless the eggs are to be used immediately. (*Chapter 7*)

Potentially hazardous foods are moist, high-protein foods on which bacteria can grow most easily. (*Chapters 1, 2*)

Punitive damages are awarded in addition to normal compensation to punish the defendant for wanton and willful neglect. (*Introduction*)

Ready-to-eat foods are properly cooked potentially hazardous foods and foods, such as vegetables or fruits, that will not be cooked. (*Chapter 1*)

Reasonable care is a possible defense against a food-related lawsuit. It is based on proving that a foodservice operation has done everything that can be reasonably expected to prevent illness by ensuring that safe food was served. (*Introduction*)

Recovery rate is the length of time it takes to produce hot water once a water heater's supply is low enough to start refilling. (*Chapter 9*)

Repellent is a liquid, powder, or mist that keeps insects away from an area but does not necessarily kill them. (*Chapter 11*)

Residual spray is used in cracks and crevices and leaves a film of insecticide that the insect absorbs or picks up as it crawls across the insecticide film. (*Chapter 11*)

Risk is the chance that a condition or a set of conditions will lead to a hazard. (*Chapter 4*)

Sanitarian, also called health official or inspector, is trained in sanitation and public health. (*Chapter 12*)

Sanitary refers to being free of harmful levels of contamination. (*Chapters 1, 10*)

Scombroid intoxication is a disease caused by the chemical histamine in tuna, bluefish, or mackerels that have been left too long in the temperature danger zone. (*Chapter 2*)

Slacking is a process used during thawing that allows food to gradually warm from frozen to unfrozen so that it cooks more evenly. (*Chapter 8*)

Solid waste includes dry, bulky trash, such as glass bottles, plastic wrappers and containers, paper bags, and cardboard boxes. (*Chapter 9*)

Source is a host, carrier, or vehicle for disease-causing micro-organisms. (*Chapter 2*)

Source reduction is decreasing the amount of material received and disposed. (*Chapter 9*)

Sous vide is a form of modified atmosphere packaging (MAP). Using this method, in which partially cooked food is in a sealed package where the oxygen has been reduced or replaced with other gases, extends a product's shelf life. (*Chapters 6, 7*)

Spore is a thick-walled protective structure produced by certain bacteria to protect their cells. Spores often survive cooking, freezing, and some sanitizing mixtures. (*Chapter 2*)

Standard, also called critical limit, is a time, temperature, or other requirement that must be met to keep a food item safe. (*Chapter 4*)

Sulfiting agent is a chemical legally used by food processors to preserve freshness and color in certain vegetables, fruits, frozen potatoes and other processed foods, and certain wines. (*Chapter 2*)

Surfactant is a substance in detergents that lessens surface tension between the detergent and the soiled surface so that detergent can penetrate and loosen soil. (*Chapter 10*)

Temperature danger zone is **40° to 140°F (4.4° to 60°C)**. Check with your local jurisdiction to find out what temperatures are accepted. The FDA's *1993 Food Code* states this zone is 41° to 140°F (5° to 60°C). Some health codes specify 45° to 140°F (7.2° to 60°C) while other codes use 40° to 140°F (4.4° to 60°C). (*Chapter 2*)

Time-temperature indicator (TTI) is a strip of liquid crystals that changes color when packaged goods reach an unsafe temperature. (*Chapter 6*)

Toxigenic refers to bacteria that produce toxins which when consumed may cause illness. (*Chapter 2*)

Toxins are poisons that are produced by micro-organisms, carried by fish, or released by plants. (*Chapter 2*)

UHT packaged refers to food that has been ultra-pasteurized (high temperature/short time) and aseptically packaged. These foods may or may not be stored under refrigeration until they are opened. After being opened, they must be refrigerated. (*Chapters 6, 7*)

Ultra-pasteurized packaged refers to food that must be refrigerated because it has been heat-treated but not necessarily aseptically packaged. (*Chapters 6, 7*)

Variance is a change or suspension from the rules of the local food code for a specific procedure or part of the restaurant granted by a regulatory agency. (*Chapter 12*)

Vegetative cells can grow and reproduce. (*Chapter 2*)

Vehicle is an item, such as wind, water, human hands, or dirty utensils, that carries or transports disease-causing micro-organisms. (*Chapter 2*)

Ventilation removes steam, smoke, grease, and heat from equipment and food preparation areas, replacing the air that was removed with clean air. (*Chapter 9*)

Virus, the smallest and simplest life-form known, is protein-wrapped genetic material. (*Chapter 2*)

Warranty of sale refers to the rules stating how the food must be handled in foodservice operations. (*Introduction*)

Water activity, expressed as A_w, is the amount of usable water (moisture) in food. Harmful bacteria will not grow at an A_w below 0.85. (*Chapter 2*)

Workflow refers to the order of the tasks to prepare a food item, beginning in the receiving area and leading to the dining room. (*Chapter 9*)

Yeasts are fungi that require sugar and moisture to survive. They spoil the foods, such as jellies and honey, in which they eat these ingredients. (*Chapter 2*)

CULINARY TERMS

Al dente, "To the tooth," is to cook pasta or vegetables until they are tender but still firm, not soft.

Blanche is to briefly cook an item in boiling water or hot fat before finishing or storing it.

Bouquet garni is a small bundle of herbs tied with string. It is used to flavor stocks, braises, and other preparations. It usually contains bay leaf, parsley, and thyme, and possibly other aromatics.

Butterfly is to horizontally cut a seafood or meat, such as chicken breast, pork chop, or tuna steak, and to open out the edges like a book or the wings of a butterfly.

Caramelize is to brown sugar in the presence of heat; the sugar may be refined or a natural component of the food, such as in an onion. Caramelization may be achieved by sautéing, broiling, or baking. The temperature range in which sugar caramelizes is approximately 320° to 360°F (160° to 182°C).

Cheesecloth is a light, fine mesh gauze used for making sachets and for straining liquids for soups, stocks, and sauces.

Chinoise, also called a China cap, is a cone-shaped sieve used to strain stocks and sauces and to purée food products.

Clarify is a process to remove solid impurities from a liquid, such as butter or stock.

Concassé usually refers to tomatoes that have been peeled, seeded, and chopped. It is used in sauces, soups, and vegetable stocks.

Dredge is to coat food with a dry ingredient, such as flour or bread crumbs.

Egg wash is a mixture of beaten eggs (whole eggs, yolks, or whites) and a liquid (usually milk or water) used to coat items that are breaded before frying.

Mirepoix is a combination of chopped aromatic vegetables (usually two parts onion, one part carrot, and one part celery) used to flavor stocks, soups, braises, and stews.

Nappé means thickened; covered with sauce.

Pincé is to caramelize an item by sautéing; usually refers to a tomato product.

Reduce is to decrease the volume of a liquid by simmering or boiling; used to provide a thicker consistency or concentration, or both.

Remouillage is a stock made from bones that have already been used for stock; it is weaker than a first-quality stock and is often reduced to make glaze.

Render is to melt fat and clarify the drippings for use in sautéing or panfrying.

Roux is a mixture containing equal parts of flour and rendered fat (usually butter) that is used to thicken sauces and soups. It can be cooked to varying degrees, such as white, pale, blond, or brown.

Sachet d'épices, "Bag of spices," refers to aromatic ingredients, encased in a cheesecloth, that are used to flavor stocks and other liquids. A standard sachet contains parsley stems, cracked peppercorns, dried thyme, and a bay leaf.

Standard breading procedure is the assembly-line procedure in which food items are dredged in flour, dipped in beaten egg, then coated with crumbs before being panfried or deep fried.

Sweat is to cook a food item, usually a vegetable, in a covered pan in a small amount of fat until it softens and releases moisture.

Truss is to tie up meat or poultry with string before cooking it in order to give it a compact shape for more even cooking and a better appearance.

ANSWER KEY

Chapter 1 Answers and Text Page References

1. a p. 7
2. c p. 8
3. b p. 11
4. d p. 11
5. a p. 8
6. c p. 10
7. d p. 8
8. c p. 8

Chapter 2 Answers and Text Page References

1. d p. 17
2. c p. 16, 20
3. d p. 16
4. d p. 20
5. a p. 16
6. b p. 23
7. c p. 23
8. d p. 25
9. c p. 26
10. d p. 27

Chapter 3 Answers and Text Page References

1. d p. 33
2. c p. 31
3. b p. 33
4. a p. 30
5. d p. 32
6. c p. 32
7. a p. 32
8. c p. 30
9. d p. 34
10. d p. 35

Chapter 4 Answers and Text Page References

1. d p. 48
2. c p. 48
3. a p. 57
4. c p. 58
5. a p. 39
6. c p. 45
7. b p. 42
8. d p. 46
9. c p. 46
10. b p. 46

Chapter 5 Answers and Text Page References

1. a p. 61
2. b p. 62
3. c p. 63
4. d p. 64
5. b p. 65
6. d p. 67
7. a p. 67
8. c p. 68
9. b p. 68
10. c p. 69

Chapter 6 Answers and Text Page References

1. a p. 72
2. c p. 80
3. b p. 73
4. b p. 83
5. b p. 82
6. d p. 85
7. d p. 82
8. c p. 85
9. c p. 74
10. c p. 86

Chapter 7 Answers and Text Page References

1. a   p. 97
2. b   p. 89
3. c   p. 90
4. d   p. 90
5. b   p. 91
6. a   p. 91
7. b   p. 92
8. d   p. 90
9. d   p. 94
10. c   p. 95

Chapter 8 Answers and Text Page References

1. d   p. 102
2. a   p. 101
3. c   p. 104
4. a   p. 104
5. b   p. 102
6. c   p. 103
7. c   p. 106
8. d   p. 107
9. a   p. 108
10. a   p. 109

Chapter 9 Answers and Text Page References

1. b   p. 114
2. b   p. 115
3. a   p. 120
4. c   p. 122
5. c   p. 118
6. d   p. 123
7. b   p. 123
8. a   p. 123
9. b   p. 126
10. c   p. 126

Chapter 10 Answers and Text Page References

1. b   p. 131
2. d   p. 137
3. b   p. 133
4. b   p. 135
5. c   p. 137
6. d   p. 140
7. a   p. 140
8. d   p. 143
9. d   p. 135
10. c   p. 141

Chapter 11 Answers and Text Page References

1. d   p. 147
2. a   p. 149
3. c   p. 150
4. b   p. 151
5. a   p. 151
6. a   p. 152
7. c   p. 156
8. c   p. 147, 153
9. b   p. 156
10. d   p. 154

Chapter 12 Answers and Text Page References

1. a   p. 160
2. b   p. 161
3. c   p. 162
4. a   p. 162
5. d   p. 162
6. c   p. 162
7. a   p. 165
8. a   p. 166
9. b   p. 166
10. a   p. 167

MORE ON THE SUBJECT

Books

American Egg Board. *A Scientist Speaks About the Microbiology of Eggs.* Park Ridge, IL: American Egg Board, revised July 1990. #E-0010.

Benenson, Abram S. *Control of Communicable Diseases in Man, Fifteenth Edition.* Washington: American Public Health Association, 1990. (512 pages.)

Bryan, F.L., C.A. Bartleson, C.O. Cook, P. Fischer, J.J. Guzewich, B.J. Humm, R.C. Swanson, and E.C.D. Todd. *Procedures to Implement the Hazard Analysis Critical Control Point System..* Ames, IA: International Association of Milk, Food and Environmental Sanitarians, 1991. (72 pages.)

The Educational Foundation of the National Restaurant Association. *HACCP Reference Book.* Chicago: The Educational Foundation of the National Restaurant Association, 1993. (198 pages.)

Miller, Jack E., and Mary Porter. *Supervision in the Hospitality Industry, Second Edition.* Chicago: The Educational Foundation of the National Restaurant Association, 1992. (347 pages.)

Miller, Jack E., and Mary Walk. *Personnel Training Manual for the Hospitality Industry.* New York: Van Nostrand Reinhold, 1991.

National Fresh Fruit and Vegetable Association. *Fresh Fruit Selection and Care and Fresh Vegetable Selection and Care.* Alexandria, VA: United Fresh Fruit and Vegetable Association, 1990. #PL1414 and #PL1418.

National Frozen Food Association. *Frozen Food Book of Knowledge—A Foodservice Reference, Tenth Edition.* Harrisburg, PA: National Frozen Food Association, Inc., 1992. (391 pages.)

Sherry, John E. *Legal Aspects of Hospitality Management. Second Edition.* Chicago: The Educational Foundation of the National Restaurant Association, 1994. (432 pages.)

Warfel, M.C., and Marion L. Cremer. *Purchasing for Foodservice Managers, Second Edition.* Berkeley, CA: McCutchan Publishing, 1990. (555 pages.)

Articles

Adams, Catherine E. "Applying HACCP to Sous Vide Products." *Food Technology* (April 1991): 148–51.

Association of Food & Drug Officials. "Guidelines for the Transportation of Food." *Journal of the Association of Food & Drug Officials* (December 1990): 85–89.

Association of Food & Drug Officials. "Retail Guidelines—Refrigerated Foods in Reduced Oxygen Packages." *Journal of the Association of Food & Drug Officials* (December 1990): 80–84.

Bean, Nancy H., and Patricia M. Griffin. "Foodborne Disease Outbreaks in the United States, 1973–87: Pathogens, Vehicles, and Trends." *Journal of Food Protection* (September 1990): 804–17.

Bryan, Frank L. "Application of HACCP to Ready-to-Eat Chilled Foods." *Food Technology* (July 1990): 70–77.

Bryan, Frank L. "Hazard Analysis Critical Control Point (HACCP) Concept." *Journal of Food Protection* (July 1990): 416–18.

Bryan, Frank L. "Hazard Analysis Critical Control Point (HACCP) Systems for Retail Food and Restaurant Operations." *Journal of Food Protection* (November 1990): 978–83.

Bryan, Frank L., P. Teufel, S. Riaz, S. Roohi, F. Qadar, and Z. Malik. "Hazards and Critical Control Points of Vending Operations at a Railway Station and a Bus Station in Pakistan." *Journal of Food Protection* (July 1992): 534–541.

Corlett, Donald A., Jr. "Regulatory Verification of Industrial HACCP Systems." *Food Technology* (April 1991): 144–46.

Curiale, Michael S. "Shelf-Life Evaluation Analysis." *Dairy, Food and Environmental Sanitation* (July 1991): 364–69.

Durocher, Joseph. "Sanitation Systems." *Restaurant Business* (May 20, 1991): 174, 176.

Gorman, J. Richard. "HACCP and Filth in Food: The Detention and Elimination of Pest Infestation." *Journal of Environmental Health* (September/October 1989): 84–86.

Herlong, Joan E. "The Power of Employee Training: These Restaurateurs Are Believers." *Restaurants USA* (August 1991): 14–17.

Indermill, Kathy. "Managing Self-Esteem." *Restaurant Business* (May 1, 1991): 118–20.

Keegan, Peter O. "Quality Training Doesn't Have to be Bankbreaking." *Nation's Restaurant News* (June 3, 1991): 56.

Lydecker, Toni. "How Self-Inspection Flies: HACCP Systems in Airline Catering Companies." *Food Service Director* (July 15, 1991): 87.

Lydecker, Toni. "Sanitation—Clean Equipment." *Food Service Director* (April 15, 1991): 113.

Lydecker, Toni. "Sanitation—Safe Salads." *Food Service Director* (March 15, 1991): 126.

Martin, Paul. "Hazard Control." *Restaurant Business* (May 1, 1991): 256.

Martin, Paul. "Managerial Changes." *Restaurant Business* (May 20, 1991): 172.

McIntyre, Charles R. "Hazard Analysis Critical Control Point (HACCP) Identification." *Dairy, Food and Environmental Sanitation* (July 1991): 357–58.

Molenda, John R. "Cholera, John Snow and the Pump Handle." *Dairy, Food and Environmental Sanitation* (January 1992): 12–15.

Molenda, John R. "Tracking Down Typhoid Mary." *Dairy, Food and Environmental Sanitation* (October 1990): 602–06.

Moore, G., T. Landreth, D. Siem, K. Sheppard, and J. Hall. "Food Sanitation Enforcement: County Inspection Program Increases Compliance." *Journal of Environmental Health* (September/October 1990): 17–18.

Patterson, Pat. "Receiving is the Front Line of Food-Quality Control." *Nation's Restaurant News* (April 22, 1991): 19, 31.

Pisciella, John A. "Overcoming the Barriers to HACCP in Restaurants." *Food Protection Inside Report* (July-August 1991): 2A.

Postel, Rose T. "A Team Manager Model for the Instruction of Quantity Food Production." *Hospitality & Tourism Educator* (November 1992): 25–29.

Restaino, Lawrence, and Charles E. Wind. "Antimicrobial Effectiveness of Handwashing for Food Establishments." *Dairy, Food and Environmental Sanitation* (March 1990): 136–41.

Rhodes, Martha E. "Educating Professionals and Consumers About Extended-Shelf-Life Refrigerated Foods." *Food Technology* (April 1991): 182–83.

Thompson, Pamela K. "Managing Teens." *Restaurant Hospitality* (May 1991): 54.

Weinstein, Jeff. "The Clean Restaurant I: Physical Plant." *Restaurants & Institutions* (May 1, 1991): 90–107.

Weinstein, Jeff. "The Clean Restaurant II: Employee Hygiene." *Restaurants & Institutions* (May 15, 1991): 138–148.

Winston, Marvin E. "Food Service Sanitation Guidelines to Avoid Food Poisoning Outbreaks." *Dairy, Food and Environmental Sanitation* (August 1991): 430–31.

Wolf, Isabel D. "Critical Issues in Food Safety, 1991–2000." *Food Technology* (January 1992): 64–70.

Videos

Close Encounters of the Bird Kind. Ternelle Productions, 1990. 18 minutes.

Commensal Rodents: Biology & Behavior and Inspection for Commensal Rodents. The National Pest Control Association, 1990. 15 minutes each.

First Impressions Last. Food Service Training Videos, 1991. 15 minutes.

Food Safety: For Goodness Sake, Keep Food Safe. Iowa State University Extension, 1991. 15 minutes.

Foodservice Disposables: Should I Feel Guilty? Foodservice & Packaging Institute, 1991. 12 minutes.

Foodservice Egg Handling & Safety. The American Egg Board, 1991. 11 minutes.

On the Front Line. National Automatic Merchandisers Association, 1990. 18 minutes.

Sanitizing for Safety—Foodborne Illness: How Can You Prevent It? Chlorox Company, 1990. 17 minutes.

Seacare: A Quality Program for Foodservice Operators. The National Fisheries Institute, 1990. 20 minutes.

Serving Safe Food Video Series, Second Edition. Introduction to Food Safety. The Educational Foundation of the National Restaurant Association, 1993. 10 minutes.

Serving Safe Food Video Series, Second Edition. Managing Food Safety—A Practical Approach to HACCP. The Educational Foundation of the National Restaurant Association, 1993. 20 minutes.

Serving Safe Food Video Series, Second Edition. Personal Hygiene. The Educational Foundation of the National Restaurant Association, 1993. 10 minutes.

Serving Safe Food Video Series, Second Edition. Preparation, Cooking, and Service. The Educational Foundation of the National Restaurant Association, 1993. 10 minutes.

Serving Safe Food Video Series, Second Edition. Proper Cleaning and Sanitizing. The Educational Foundation of the National Restaurant Association, 1993. 10 minutes.

Serving Safe Food Video Series, Second Edition. Receiving and Storage. The Educational Foundation of the National Restaurant Association, 1993. 10 minutes.

DIRECTORY OF FOOD SAFETY RESOURCES

American Egg Board
Foodservice Manager
1460 Renaissance Dr.
Park Ridge, IL 60068

American Public Health Association (APHA)
1015 Fifteenth St. NW
Washington, DC 20005

The Association of Food and Drug Officials (AFDO)
P.O. Box 3425
York, PA 17402

Beef Industry Council of the National Live Stock and Meat Board
444 N. Michigan Ave.
Chicago, IL 60611

Centers for Disease Control and Prevention (CDC)
Public Inquiries Specialist
1600 Clifton Road, NE
Atlanta, GA 30333

The Educational Foundation of the National Restaurant Association
Technical Education Department
250 S. Wacker Dr.
Ste. 1400
Chicago, IL 60606-5834

Foodservice and Packaging Institute
1025 Connecticut Ave., NW
Ste. 513
Washington, DC 20036

International Association of Milk, Food and Environmental Sanitarians (IAMFES)
200W Merle Hay Centre
6200 Aurora Ave.
Des Moines, IA 50322

International Council of Hotels and Restaurant Industry Educators
(International CHRIE)
1200 17th St., NW, 7th Fl.
Washington, DC 20036

International Food Manufacturers Association (IFMA)
321 N. Clark Street
Ste 2900
Chicago, IL 60610

National Automatic Merchandisers Association (NAMA)
20 N. Wacker Dr., 35th Fl.
Chicago, IL 60606

The National Environmental Health Association (NEHA)
720 S. Colorado Blvd.
Ste. 970
Denver, CO 80222

National Fisheries Institute (NFI)
1525 Wilson Blvd.
Ste. 500
Arlington, VA 22209

National Food Processors Association (NFPA)
1401 New York Ave. NW
Washington, DC 20005

National Frozen Food Association
P.O. Box 6069
Harrisburg, PA 17112

National Live Stock and Meat Board
444 N. Michigan Ave.
Chicago, IL 60611

The National Pest Control Association (NPCA)
8100 Oak St.
Dunn Loring, VA 22027

The National Restaurant Association
Technical Services or Information Services and Library
1200 17th St., N.W.
Washington, DC 20036-3097

NSF *International*
3475 Plymouth Road
P.O. Box 1468
Ann Arbor, MI 48106

Soap and Detergent Association
475 Park Avenue South
New York, NY 10016

U.S. Department of Agriculture (USDA)
Food and Nutrition Service
50 E. Washington St.
Chicago, IL 60602
   Meat and Poultry Hotline
   800/535–4555

U.S. Food and Drug Administration (FDA)
Retail Food Protection Branch
HFF-342
200 C. St. NW
Washington, DC 20204

INDEX

A

Abrasive cleaner, 132
Acceptable and Unacceptable Conditions for Receiving Meat, Poultry, Fish, and Seafood, *chart,* 83
Acid cleaner, 132
Acidity, 17
ADA. *See* Americans with Disabilities Act
Additive, 25
Air curtain, *or* air door, *or* fly fan, 148, *illus.,* 148
Air gap, 123, *diagrams,* 124, 125
Albumen, *chart,* 18
Alfalfa sprouts, 8, *illus.,* 9
Alkalinity, 17
American cockroach (pest), *chart,* 151
American Society of Sanitary Engineers, *or* ASSE, 123
Americans with Disabilities Act, *or* ADA, 35
Anisakiasis, 23
Anisakis (parasite), 23
Ant (pest), 152
Aseptic packaging, 85, 96
Asian cockroach (pest), *chart,* 151
A$_w$. *See* Water activity

B

Bacillary dysentery. *See* Shigellosis
Bacillus cereus (bacteria), 20, *chart,* 22
Bacillus cereus gastroenteritis, *chart,* 22
Backflow, 123
Back-siphonage, 123, *illus.,* 124
Bacon
    cooking, *chart,* 103
    water activity level, *diagram,* 19
    See also: Appendix C
Bacteria (singular, bacterium), 15, *illus.,* 10
    cutting board, 120, *illus.,* 121
    reproduction, *illus.,* 16
    Salmonella temperature and growth, *illus.,* 18
    water activity, *diagram,* 19
    See also: specific types by name, *e.g., Campylobacter jejuni*
Bacterium. *See* Bacteria
Batch-type dishwasher, 121
Bathing, 33
Batter, 105
Beef
    cooking, *charts,* 103
    meat and poultry grades, 72, *illus.,* 73
    potentially hazardous foods, 8, *illus.,* 9
    See also: Appendix C; Appendix D
Beetle (pest), 152
Bi-metallic stemmed thermometer, 77, *illus.,* 78
Biological hazards, 15
    contamination and foodborne illness, 8, *illus.,* 10
Bird (pest), 153
Blast chiller, 119
    refrigerated storage, 108, *illus.,* 109, 119
Boiling point method, 79

Botulism
    canned foods inspection, 85
    foodborne illness, *chart,* 22
Breading, 105
Brown-banded cockroach (pest), *chart,* 151
Buffet, 63
Butter, 82
    See also: Appendix C; Appendix D

C

Cadmium, 26
Cafeteria, 63
Calibrating (thermometer), 79, *illus.,* 79
Campylobacteriosis, *chart,* 22
Campylobacter jejuni (bacteria), *chart,* 22
Candy thermometer, 78
Canned food
    receiving, 85
    storing, 96
    See also: Appendix E
Cantilever mounted equipment, 122
Carousel dishwasher, 121
Carrier, 20
Catering, 67
CCP. *See* Critical control point
Ceiling
    cleaning and sanitizing, 139
    sanitary facility design, 116
Central kitchen, 65
Ceramic tile, 116
CFR, *or* Code of Federal Regulations, 161
Cheese
    receiving procedures, 82
    water activity level, *diagram,* 19
    See also: Appendix C; Appendix D
Chemical. *See* Foodservice chemicals
Chemical hazards, 24
    contamination and foodborne illness, 10, *illus.,* 10
Chemical-sanitizing, 132
    dishwashing equipment, 120, 135, *chart,* 136
    guidelines, *chart,* 134
Chicken
    HACCP system, 42
    pH level, *diagram,* 18
    water activity, *diagram,* 19
    See also: Poultry; Appendix C
Children, 8, 42
Chili Flowchart, *illus.,* 44
Chili Recipe, *illus.,* 41
    HACCP system, *chart,* 49
Chili Recipe with CCPs, *illus.,* 43
Chlorine, 133, *charts,* 134, 136
Ciguatera, 24
Ciguatoxin, 24
Cinder block, 116

Escherichia coli 0157:H7 (bacteria)
    employee illness, 35
    foodborne illness, 20, *chart,* 22
Escherichia coli 0157:H7 enteritis, *chart,* 22

F

Facility
    cleaning and sanitizing, 139
    food safety, 80
    integrated pest management, 148
    layout and work flow design, 113
    storage equipment, 96
    See also: Foodservice operation
Factors Most Often Named in Foodborne Outbreaks, *list,* 12
Factors That Influence Chemical Sanitizers, *chart,* 133
Factors That Influence Cleaning, *chart,* 131
FAT-TOM, 17
FDA. *See* Food and Drug Administration
Fingernails, 33
First In, First Out, *or* FIFO
    storage, 89
Fish
    cooking, *chart,* 103
    poisonous, 24
    potentially hazardous food, 8, *illus.,* 9
    purchasing, 74
    receiving, *illus.,* 83
    storage, 94
    See also: Appendix C; Appendix D
Flight-type dishwasher, 121
Flooring
    cleaning and sanitizing, 139
    facility design, 115, *illus.,* 116
    integrated pest management, 149
Flour. *See* Appendix E
Flowchart, 45, *chart,* 44
Flow of food, 9
    HACCP system, 40
Fly (pest), 151
Fly fan. *See* Air curtain
Food
    bacteria reproduction, 16
    contamination and foodborne illness, 3, 17, 20, *chart,* 21
    delivery. *See* Receiving
    dry. *See* Dry food
    employee tasting, 34
    FAT-TOM, 17
    frozen. *See* Frozen food
    pH of some common foods, *diagram,* 18
    shipments. *See* Inspection; Rejecting shipments
    tasting, 34
    water activity of some common foods, *diagram,* 19
    See also: Appendix A; Appendix B; Appendix C; Appendix D;
        Appendix E
Food and Drug Administration, *or* FDA, 5
    cooling procedures, 108, *illus.,* 109
    Food Code (1993), 17, 31, 35, 95, 108, 161, 165
    regulatory agency, 161
Food bar, 63
Foodborne illness, 7

bacteria, 20
causes, 12
contamination, 17
customers, 3
    major diseases, *chart,* 21
Foodborne infection, 20, *chart,* 21
Foodborne intoxication, 20, *chart,* 21
Foodborne outbreak. *See* Outbreak
Food-contact surface, 11
    cleaning and sanitizing, 130
Food safety
    HACCP system, 6, 39
    hazard control safety program, 48
    preparation and service, 100
    ten rules, 110
Foodservice chemicals, 26
    See also: specific chemicals by name, *e.g.,* Chlorine; Iodine
Foodservice operation
    HACCP system, 42
Freezer, 117
Freezing, 92
Frozen food
    receiving, 85
    storage, 92
    See also: Appendix D
Fruit
    potentially hazardous foods, 8
    See also: Produce; Appendix A; Appendix C; Appendix D;
        Appendix E
Full-service operation, 62
Fungi, 23, *illus.,* 10

G

Galvanized containers, 26
Game (wild) meat
    cooking, *chart,* 103
    purchasing, 72
Garbage, 126
    integrated pest management, 149
Garbage Disposal, *illus.,* 127
Garlic-and-oil mixture
    potentially hazardous foods, 8, *illus.,* 9
General Guidelines for Chemical Sanitizers, *chart,* 134
Generally Recognized As Safe, *or* GRAS, 25
German cockroach (pest), *chart,* 151
Gloves, 33
Glue board, 156
Glue trap, 151
Government regulatory agency, 160
GRAS. *See* Generally Recognized As Safe
Group training, 58
Gum-chewing, 32

H

HACCP. *See* Hazard Analysis Critical Control Point
HACCP System for Chili, *diagram,* 49
Hair, 33